T0214711

Problem Books in Mathematics

Series Editor:

Peter Winkler
Department of Mathematics
Dartmouth College
Hanover, NH 03755
USA

More information about this series at http://www.springer.com/series/714

Kseniya Garaschuk • Andy Liu

Grade Five Competition from the Leningrad Mathematical Olympiad

1979–1992

 Springer

Kseniya Garaschuk
Department of Mathematics and Statistics
University of the Fraser Valley
Abbotsford, BC, Canada

Andy Liu
Department of Mathematical
and Statistical Sciences
University of Alberta
Edmonton, AB, Canada

ISSN 0941-3502   ISSN 2197-8506   (electronic)
Problem Books in Mathematics
ISBN 978-3-030-52948-2   ISBN 978-3-030-52946-8   (eBook)
https://doi.org/10.1007/978-3-030-52946-8

Mathematics Subject Classification (2020): 00A07, 00A09, 97-XX

This Springer imprint is published by the registered company Springer Nature Switzerland AG
The registered company address is: Gewerbestrasse 11, 6330 Cham, Switzerland

Preface

The Leningrad Mathematical Olympiad was the oldest mathematics competition in the former Soviet Union. It began in the spring of 1934, predating the Tbilisi Mathematical Olympiad by a few months and the Moscow Mathematical Olympiad by one year.

It was organized as a three-stage contest. The first (School level) round was quite informal and often the teachers skipped the selection, simply telling the willing students they should go and take part in the second round. However, the teachers were supplied with the lists of training and selection problems distributed by the local district instructors.

The second (District level) round was a regular formal written contest, held at several locations around each city district. The number of districts changed with times, but in the 1980s there were fifteen of them, serving approximately 200,000 of middle and high school students.

Every year, about 10,000 of these school kids took part in the first two rounds of the Leningrad Mathematical Olympiad. A much smaller number of the second round winners (between 80 and 150 per each grade) were invited to the third (All-city) round. This competition was (and still is) unique for the following two reasons.

First, the majority of the problems used at the Leningrad Mathematical Olympiad would be designed specifically for this competition. Almost all of them were never used in other competitions, nor are they borrowed from the existing compilations. That is not an easy feat which requires quite a lot of work and ingenuity from the problem committee.

Second, it was also unique in that an *oral* element was involved. It has often been referred to as an oral competition, but this is, strictly speaking, not correct. A truly oral competition would encounter insurmountable logistic and pedagogical difficulties.

It was the *evaluation* of the Leningrad Mathematical Olympiad that took an oral form. The competition started in the traditional setting, with the contestants assembled in a classroom. The competition papers would be distributed to them, and they would have pencils and paper to work on the problems.

At a moment chosen by an individual contestant, say Natasha, she would raise her hand, indicating that she believed she had solved a problem, and would like her work evaluated. An invigilator would take her out of the classroom into an area where they would not be disturbed. There, she would give an oral presentation of her proposed solution.

A + or − would be assessed. If the assessment was +, Natasha would receive full credit for the problem. She would return to the classroom if the global time limit for the competition had not been exceeded. She might choose to continue to work on the same problem if the assessment was −. However, she could only present her solution to the same problem at most three times. If in the end the problem is solved, the erroneous attempts are discounted, with the final result of a contestant being simply the total number of problems solved.

This system was inspired by the so-called *mathematical circles*, a type of informal seminars in extracurricular mathematics held by young professors and students of the Leningrad universities in the beginning of 1930s. In turn, these seminars have mimicked the regular flow of some university lectures, where a few professors permitted (actually, requested!) the students to interrupt their explanations and proofs with questions thus allowing for a quick and often enlightening exchange of ideas right in the middle of a lecture. As far as we know, none of the official mathematical competitions around the world use (or ever used) this system.

The oral system has several significant advantages over the usual written competition. First, it allows for a quick fix of a minor mistake or a typo. Second, it allows for an extra attempt. Third, it often introduces the students to a different type of thinking during their back-and-forth exchange with the invigilators. It has its drawbacks, too. One of them is that the invigilators can make mistakes, too. Second is a major logistical hurdle in organizing an oral event on such a scale. Indeed, on the day of the Olympiad, up to 400 students from three grades (middle or high school) would attend. This required a lot of rooms and a lot of helpers. In Leningrad that problem was resolved due to a lot of cooperation the Olympiad received from the two major universities; namely, Leningrad State University (the only full-scale city university at the time) and Leningrad Pedagogical Institute. These institutions permitted the use of their campuses and their classrooms.

Originally, the competition was held for the high-school students only, with one common set of questions. In the later years, separate grade-specific approach was introduced, and so the problem sets had to designed for each grade. Of course, one grade's set could "borrow" from another. By the end of 1950s all Grades Six through Ten had their "own" Olympiad. The fifth grade had to wait until 1969. However, the Grade Five Olympiad was entrusted to a different committee which organized that event for several years. Alas, its records were not recovered and we are not in possession of the problems sets before 1979.

At the time, the Soviet school system spanned ten years, whereas the current Canadian school system spans twelve years. Publicly, we would puff up our cheeks and declare that our Grade Seven is equivalent to their Grade Five. Privately, we would be jumping for joy if our Grade Nine students can handle this Grade Five competition.

Nevertheless, the former Soviet students in Grade Five were still at a very tender age. What sort of meaningful questions could one ask at that level? A quick scan of this book is eye-opening. We have always had a healthy respect for the intellectual abilities of children, but their imagination could be stretched much further.

In 1990, a year was added to the Soviet school system. We have decided to cut our coverage off in 1992, the year immediately after the dissolution of the former Soviet Union. In particular, Leningrad reverted to its original name, St. Petersburg. So that seems a natural breaking point. Therefore, for the period 1990 – 1992, we present what had become the Grade Six Competition still under the banner of the Grade Five Competition.

We pay tribute to all the undergraduate and graduate students as well as professors of post-secondary institutions in Leningrad who generously donated their time and efforts working as invigilators. Without them, the Leningrad Mathematics Olympiad would not happen. Most important of all, we are grateful to the composers of the wonderful contest problems for their creative genius. Unfortunately, the record is far from complete. What information we do possess is in the table below.

Proposer	Contest Problems (Year/Number)
E. V. Abakumov	88.06
A. V. Bogomol'naya	88.01, 88.04
D. V. Fomin	84.05, 85.03, 88.02, 88.03, 89.02, 89.03, 89.04, 90.01, 90.02, 90.03, 90.05, 91.02, 91.03, 91.05, 91.06, 92.05
S. V. Fomin	80.03, 80.05, 82.01, 82.02, 82.04, 82.06, 83.03, 84.02, 85.01, 85.04, 85.05, 86.03, 86.06, 87.02, 87.04, 87.06
S. A. Genkin	84.01, 84.03, 84.06, 86.01, 86.02, 87.01, 87.05, 88.05, 89.01, 89.06
M. N. Gusarov	89.05
I. V. Itenberg	84.04, 90.04, 91.06
K. P. Kokhas	92.01, 91.01
F. L. Nazarov	90.06, 91.04, 92.04, 92.06
A .E. Perlin	92.02, 92.03

Dmitri Fomin, the most prolific composer among this elite group, has written a book in Russian while he was still with the St. Petersburg State University. *St. Petersburg Mathematical Olympiads* was published by Politechnica Publishing of St. Petersburg in 1994, with 309 pages under the ISBN 5-7325-0363-3. Dmitri has since emigrated to the United States and is currently working in the private sector in Boston. He is preparing an English translation of his book. We thank him for sharing with us details about the Olympiad in the olden golden days.

Paul Vaderlind of Stockholm has provided us a photocopy of the Russian book. We have translated the Grade Five problems and worked out our own solutions. The Chapter titles are taken from the famous four-step method in problem-solving of the great Hungarian mathematics educator **György Pólya**. We hope that our effort has enhanced the value of what is already an amazing legacy.

The competition problems are given chronologically in Chapter One. In Chapter Two, the eighty-three problems are divided loosely into twenty-six sets of three or four related problems. For each, an example is provided. Often, it is a simplification of the actual problem, but in any case, the example is chosen to provide assistance in solving the problem. In Chapter Three, full solutions to all the problems are given. In Chapter Four, generalizations of the problems and further investigation based on them are explored.

We thank Vladimir Troitsky of the University of Alberta who has provided invaluable help during the preparation of this book. We are grateful to Robinson dos Santos, Anne Comment, Jan Holland, Saveetha Balasundaram, Gomathi Mohanarangan and Jeffrey Taub of Springer Nature for their encouragement, advice and support.

Kseniya Garaschuk,
Abbotsford, BC, Canada.
Andy Liu,
Edmonton, AB, Canada.
2020.

Table of Contents

Detailed Table of Contents

Chapter One: Understand the Problem

1979

1. Pack fifteen 2×3 chocolate pieces into a 7×13 box, leaving a 1×1 hole.

2. In 1979, Natasha's age was equal to the sum of the digits of the year when she was born. What year was that?

3. Counters are placed on 25 of the squares of a 6×7 chessboard. Prove that there exists a 2×2 subboard with at least three counters on its four squares.

4. Each of the digits 0 to 9 is written on a card.

   (a) Prove that from any three of the ten cards, a multiple of 3 with up to three digits can be formed.
   (b) What is the minimum number of cards from which a multiple of 9 with up to nine digits can always be formed?

5. A class consists of 31 grade 2 students and some grade 3 students. There are 19 tables at which one or two students may sit. If each boy knows exactly three girls and each girl knows exactly two boys, how many students are there altogether?

© The Editor(s) (if applicable) and The Author(s), under exclusive license to Springer Nature Switzerland AG 2020
K. Garaschuk, A. Liu, *Grade Five Competition from the Leningrad Mathematical Olympiad*,
Problem Books in Mathematics, https://doi.org/10.1007/978-3-030-52946-8_1

1. Is it possible to construct a 5×6 table with the integers from 1 to 30 such that the sum of the six numbers in each row is constant, and the sum of the five numbers in each column is also constant?

2. Each of 23 students is 10, 11, 12 or 13 years old, with at least one of each age. Their total age is 253 years. How many 12 year olds are there if there are 1.5 times as many 12 year olds as 13 year olds?

3. On the line AB, 200 points are chosen symmetrically with respect to the midpoint of AB. Half of the points are red, and half are blue. Prove that the sum of the distances from A to all the red points is equal to the sum of the distances from B to all the blue points.

4. Seven genuine coins have the same weight. Two counterfeit coins also have the same weight. A counterfeit coin is heavier than a genuine coin. Identify the fake coins using a standard balance at most four times.

5. Dissect a square into convex pentagons.

6. The map of a subway system is a convex polygon in which no three diagonals are concurrent. There is a station at each vertex and at every intersection of two diagonals. Train runs along entire diagonals, but not necessarily every diagonal. If each station lies on the route of at least one train, prove that it is possible to go from any station to any other station, changing trains at most twice.

1. Adam and Betty wrote 54 tests marked out of 3. Checking their records, Adam had as many 3s as Betty had 2s, as many 2s as Betty had 1s, and as many 1s as Betty had 0s. Prove that their averages are not the same.

2. Does there exist a number consisting of each of the digits 1 to 9 exactly once such that between any two digits differing by 1, there are an odd number of other digits?

3. Place nine points inside a triangle so that there are twelve points altogether. In addition to the three sides of the triangle, join some other pairs of these twelve points with non-intersecting segments so that each point is joined to five others, and the original triangle is divided into triangles.

4. What is the smallest number of counters that must be placed on the squares of a 12 × 12 chessboard so that the shape in Figure 1 cannot be placed on three unoccupied squares of the chessboard?

Figure 1

5. Does there exist a positive integer such that its square begins with 123456789?

6. A certain country has only four kinds of bills, worth $1, $2, $5 and $10. Prove that from a stack of bills totaling $400, an outsider can be paid exactly $300.

3

1. A six-digit number is given. How many different seven-digit numbers have the property that if one digit is removed from it, the given six-digit number will be obtained?

2. A grasshopper jumps 1 cm. Then it jumps 3 cm in the same or the opposite direction. Then it jumps 5 cm in the same or the opposite direction, and so on. Can the grasshopper get back to the starting point on the 25th jump?

3. The squares of a 5 × 5 chessboard are painted in one of two colors in an arbitrary way. Prove that there exist 2 rows and 2 columns such that the 4 squares where they intersect are all of the same color.

4. Prove that if the sum of two positive integers is 770, then their product is not divisible by 770.

5. Label the edges of a cube with 1, 2, 3, ..., 12 so that the sum of the labels of the four edges of each of the six faces is the same.

6. Anna and Boris play a game with a stack of 100 counters. Anna goes first, and moves alternate thereafter. In each move, a player divides a stack of at least two counters into two smaller piles. The loser is the player without a move, when each stack consists of exactly one counter. Prove that Anna must win this game.

1. In a chess tournament, each participant plays every other participants exactly once. Each participant gets 1 point for a win, $\frac{1}{2}$ point for a draw and 0 points for a loss. At most how many of the 30 participants can score 18 points or more?

2. Ten 10×20 chocolate pieces are cut up into twenty triangles. Pack them into a square box, leaving no empty space.

3. Each of Benny, Denny, Kenny and Lenny either always lies or always tells the truth. Benny claims that Denny is a liar. Lenny claims that Benny is a liar. Kenny claims that both Benny and Denny are liars. Kenny also claims that Lenny is a liar. Which of them always lies and which of them always tells the truth?

4. A solitaire game starts with eight numbers arranged in a circle. Each is either 1 or -1, and the choice is arbitrary. In each move, one can multiply any blocks of adjacent numbers of length 3 by -1. Prove that one can make all eight numbers equal to 1.

5. Baron Münchhausen has a time machine which allows a jump from March 1st to November 1st of any other year, from April 1st to December 1st, from May 1st to January 1st and so on. He cannot arrive and then depart on the same day. The Baron starts and ends his time travel on April 1st, and claims that he has been away for 26 months. Prove that he is mistaken.

6. Four different digits are given. We use each of them exactly once to construct the largest possible four-digit number, and use each of them exactly once to construct the smallest possible four-digit number which does not start with 0. If the sum of these two numbers is 10477, what are the given digits?

1. Prove that in the 400-digit number 84198419...8419 some digits can be deleted from the beginning and some from the end such that the sum of the remaining digits is equal to 1984.

2. Construct a 4×4 table of non-zero numbers such that the sum of the numbers in the four corner squares of any 2×2, 3×3 or 4×4 subtable is 0.

3. On a line containing a segment AB, there are 45 points marked, none of which lie on the segment AB. Prove that the sum of the distances from these points to A is not equal the sum of their distances to B.

4. Each square of an infinite chessboard is painted in one of 8 colors. Prove that it is possible to place a copy of the shape in Figure 2, possibly rotated or reflected, such that it covers two squares of the same color.

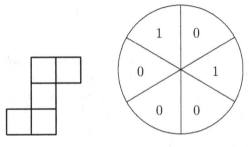

Figure 2   **Figure 3**

5. A solitaire game starts with a circle divided into six sectors each containing a number as shown in Figure 3. In each move, one may add 1 to the numbers in any two adjacent sectors. Prove that it is impossible to make all six numbers equal.

6. Anna and Boris play a game with the numbers from 1 to 100 written in order in a row. Anna goes first, and turns alternate thereafter. In each move, a player puts one of the operation signs $+$, $-$ and $\times$ between any two numbers which do not already have an operation sign in between them. After 99 operation signs have been placed, the value of the expression is computed. Anna wins if this value is odd, and Boris wins if it is even. Prove that Anna has a winning strategy.

1. Each of 68 coins has a different weight. In 100 weighings on a standard balance, find the heaviest and lightest coins.

2. A 45-digit number consists of one 1, two 2s, three 3s, ..., and nine 9s. Prove that it is not the square of an integer.

3. A traveler departed from his home city A to the city B farthest from A. From B, he departed to the city C farthest from B, and so on. Prove that if C and A are different cities, then the traveler will never reach home.

4. Find 1000 numbers whose sum is equal to their product.

5. Among 300 boots, 100 are of size 8, 100 of size 9 and 100 of size 10. There are 150 right boots and 150 left boots. Prove that one can select at least 50 pairs of boots each consisting of a right boot and a left boot of the same size.

6. A solitaire game starts with ten boxes numbered from 1 to 10 arranged randomly in two stacks. In each move, one may take several boxes from the top of a stack and put them on top of the other stack, which may be empty. Prove that one can form a stack with the ten boxes ordered from 1 at bottom to 10 at the top in 19 moves.

1. A solitaire game starts with the cards numbered 7, 8, 9, 4, 5, 6, 1, 2 and 3 placed in a row in that order. In each move, one may remove a block of any numbers of adjacent cards, reverse their order and put them back in the row. Show a sequence of moves which puts the cards in the order 1, 2, 3, 4, 5, 6, 7, 8 and 9.

2. There are 44 queens on a standard chessboard. Prove that each queen attacks at least one other queen.

3. Let a and b be positive integers such that $34a = 43b$. Prove that $a + b$ is composite.

4. Find a configuration of several identical round coins on a table so that each coin touches exactly three other coins.

5. There are 55 numbers placed on a circle. Each number is the sum of its neighbors. Prove that all numbers are equal to zero.

6. (a) Find a seven-digit number with distinct digits which is divisible by all of its digits.

   (b) Does their exist an eight-digit number with this property?

1. A solitaire game starts with the 4×4 table in Figure 4 on the left. In each move, one may add 1 to all the numbers in any row or subtract 1 from all the numbers in any column. How can one obtain the table in Figure 4 on the right?

1	2	3	4
5	6	7	8
9	10	11	12
13	14	15	16

1	5	9	13
2	6	10	14
3	7	11	15
4	8	12	16

Figure 4

2. A certain country has only four kinds of bills, worth $1, $10, $100 and $1000. Can exactly half a million bills be worth exactly one million dollars?

3. Each of six cities is connected to the others by five roads. Show that it is possible for the roads to intersect only three times with exactly two roads crossing over at each intersection. Junctions at the cities are not considered intersections.

4. The cost of a hotdog and the cost of a burger are both integral numbers of cents. If each boy buys a hotdog and each girl buys a burger, their total expenditure will be one cent more than if each boy buys a burger and each girl buys a hotdog. There are more boys than girls. How many more?

5. A six-digit number from 000000 to 999999 is said to be *lucky* if the sum of its first three digits is equal to the sum of its last three digits. What is the minimum length of a block of consecutive numbers which will guarantee the inclusion of at least one lucky number, whatever the first number of the block may be?

6. Anna and Boris play a game on a 9×9 chessboard. Anna goes first and turns alternate thereafter. In each move, Anna puts a red counter on a vacant square while Boris puts a blue counter on a vacant square. When the board is completely filled, a row with more red counters than blue counters is called a red row, and a blue row otherwise. Red and blue columns are similarly defined. The score for Anna is the sum of the numbers of red rows and red columns while that for Boris is the sum of the numbers of blue rows and blue columns. What is the highest possible score for Anna?

1. A solitaire game starts with a 0 in each square of a 3×3 table. In each move, one may add 1 to all numbers in any of the four 2×2 subtables. Is it possible to obtain the table in Figure 5?

4	9	5
10	18	12
6	13	7

Figure 5

2. A teacher has a deck of 30 cards numbered from 1 to 30, as does each of 30 students. They all turn over the top cards in their respective decks. Whenever the number on a student's card matches the number on the teacher's card, that student scores 1 point. When all the cards have been turned over, each student scores a different number of points. Prove that one of them scores 30 points.

3. Is it possible to arrange the positive integers from 1 to 100 inclusive in a row so that the difference between any two adjacent numbers is at least 50?

4. Do there exist two non-zero integers such that one of them is divisible by their sum and the other is divisible by their difference?

5. A solitaire game starts with a pile of 1001 counters. In each move, a pile with at least three counters is chosen. One counter is removed while the remaining counters are split into two piles, not necessarily equal in size. Is it possible that after a number of moves, each remaining pile contains exactly three counters?

6. A solitaire game starts with an 8×8 chessboard in which all 64 squares are white. In each move, one may enter the chessboard on a border square, visit parts of the chessboard going between two squares with a common side, and exit via a border square. The color of a square is changed from white or black or vice versa every time it is visited. Is it possible to create the usual alternating color pattern on the chessboard?

1. Each of the competition for the fifth, sixth, seventh, eighth, ninth and tenth grades consists of seven problems. In each competition, exactly four of the questions do not appear on the competition of any other grade. What is the greatest number of distinct problems that may appear in these six competitions?

2. Prove that among the six-digit numbers from 000000 to 999999, there are as many with digit sum 27 as those where the sum of its first three digits is equal to the sum of its last three digits.

3. In a model railway set, there are two kinds of track pieces as shown in Figure 6. They may not be flipped over. When two pieces are put together, the convex head of one must fit into the concave tail of the other. A closed track constructed according to this rule has been taken apart, and one piece has been replaced by a piece of the other kind. Prove that it is now impossible to reassemble the pieces to form a closed track which follows the rule.

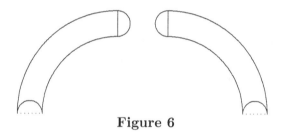

Figure 6

4. Each of 32 coins has a different weight. In 35 weighings on a standard balance, find the heaviest and the second heaviest coins.

5. Find two six-digit numbers such that the twelve-digit number obtained when the digits of one are written after the digits of the other is divisible by the product of the original six-digit numbers.

6. Anna and Boris play a game on a 10×10 chessboard. Anna goes first, and turns alternate thereafter. In each move, a player puts either a red counter or a green counter on a vacant square. A player wins by completing a block of three adjacent counters of the same color along a row, a column or a diagonal. Which player, if either, has a winning strategy?

1. Paula numbers the 96 sheets in her notebook page by page in order from 1 to 192. Nick rips out 25 sheets at random and adds together all 50 page numbers. Prove that his sum cannot be equal to 1990.

2. Among 101 coins, 100 are genuine and have the same weight. The weight of the counterfeit coin is different from that of a genuine coin. Determine whether the counterfeit coin is heavier or lighter in two weighings on a standard balance. The identification of the counterfeit coin is not required.

3. Is it possible to pack thirty-nine 5×11 chocolate pieces into a 39×55 box?

4. Anna and Boris play a game starting with the number 1234. Anna goes first, and turns alternate thereafter. In each turn, the player subtracts from the number one of its non-zero digits. A player wins if the number is reduced to 0. Who has a winning strategy, Anna or Boris?

5. Together, three students solved exactly 100 problems. Separately, each solved exactly 60 problems. A problem was considered difficult if it was solved by only one of them, and easy if it was solved by all of them. Prove that the number of difficult problems exceeded the number of easy problems by 20.

6. For any boy in a club, all the girls who know him know one another. Each girl knows more boys than girls, not counting themselves. Prove that there are at least as many boys as girls in this club.

1. Each of 40 children in a workshop class has nails, nuts and bolts. There are exactly 15 children with unequal numbers of nails and nuts, and 10 children with equal numbers of nuts and bolts. Prove that there are at least 15 children with unequal numbers of nails and bolts.

2. A strange rule in a club allows the children only to trade any two marbles for any three others, or any three marbles for any two others. Is it possible to start with 100 red marbles and end up with 100 green marbles, having traded away exactly 1991 marbles in the meantime?

3. Four girls are starting simultaneously from the same point on a circular track, running at constant speeds which are not necessarily the same. A and B run clockwise while C and D run counterclockwise. A meets C for the first time at the same moment as B meets D for the first time. Prove that A catches up with B for the first time at the same moment as D catches up with C for the first time.

4. Baron Münchhausen hunts ducks every day. One day, he declares, "Today, I will bring home more ducks than two days ago but fewer than one week ago." For at most how many consecutive days can the Baron say this without telling a lie?

5. Anna and Boris play a game with a red stick, a white stick and a blue stick, each of which is 1 meter long. Anna starts by breaking the red stick into three pieces. Then Boris breaks the white stick into three pieces. Finally, Anna breaks the blue stick into three pieces. She wins if she can use the nine pieces to form three triangles with sides of different colors. Can Boris stop her from winning?

6. In a tournament without draws, every two of the nine teams play against each other exactly once. Must there always be two teams such that every other team has lost to either or both of them?

1. In a tournament, each participant plays every other participants exactly once. Each participant gets 1 point for a win, 0 points for a drawn, and −1 point for a loss. One of the participants finishes the tournament with 7 points and another with 20. Prove that there is at least one drawn game.

2. In a heptagonal castle, each of the seven sides is a straight wall and there is a watchtower at each of the seven vertices. The guards stay in the watchtowers. Each guard watches over both walls meeting at that watchtower. What is the minimum number of guards required so that each wall is watched over by at least seven guards?

3. Adam and Betty are of the same age. Adam multiplies his age this year by his age last year. Betty calculates the square of her age next year. Prove that the two answers have different digit-sums.

4. Fyodor collects coins. No coin in his collection is more than 10 cm in diameter. He keeps all the coins arranged side by side in a rectangular box of size 30 cm by 70 cm. Prove that he can fit all of his coins in another rectangular box of size 40 cm by 60 cm.

5. A circle is divided into 27 equal arcs by 27 points. Each point is either white or black. No two black points are adjacent or separated by only one white point. Prove that three of the white points are the vertices of an equilateral triangle.

6. Three counterfeiters print bills of arbitrary integral denominations. Each one prints bills totaling $100, and can pay either of the other two counterfeiters any amount up to $25, perhaps with change. Prove that together they can pay an outsider exactly any amount from $100 to $200.

Chapter Two: Make a Plan

Set A : Problems on Differences

1. (1988-3)

   Is it possible to arrange the positive integers from 1 to 100 inclusive in a row so that the difference between any two adjacent numbers is at least 50?

   Example.

   Arranged around a circle are 10 distinct integers such that the difference between any two adjacent numbers is 2 or 3. If the smallest number is 0, what is the maximum value of the largest numbers?

   Solution:

   The largest number is at most $5 \times 3 = 15$, if it is diametrically opposite to 0, and all differences between adjacent pairs are 3. However, this means that the 10 numbers will not be distinct. Hence the largest is at most 14, and this can be attained with 0, 3, 6, 9, 12, 14, 11, 8, 5 and 2.

2. (1986-5)

   There are 55 numbers placed on a circle. Each number is the sum of its neighbors. Prove that all numbers are equal to zero.

   Example.

   There are 12 numbers placed on a circle. Each number is the sum of its neighbors. Is it necessarily true that all numbers are equal to zero?

   Solution:

   It is not true. The numbers may be 2, 1, −1, −2, −1, 1, 2, 1, −1, −2, −1 and 1 in cyclic order.

3. (1990-5)

   Together, three students have solved exactly 100 problems. Separately, each has solved exactly 60 problems. A problem is considered difficult if it is solved by only one of them, and easy if it is solved by all of them. Prove that the number of difficult problems exceeds the number of easy problems by 20.

   Example.

   Together, two students solved exactly 100 problems. Separately, each solved exactly 60 problems. A problem was considered difficult if it was solved by only one of them, and easy if it was solved by both of them. What was the difference between the number of difficult problems and the number of easy problems?

© The Editor(s) (if applicable) and The Author(s), under exclusive license to Springer Nature Switzerland AG 2020
K. Garaschuk, A. Liu, *Grade Five Competition from the Leningrad Mathematical Olympiad*,
Problem Books in Mathematics, https://doi.org/10.1007/978-3-030-52946-8_2

Solution:

Since $2 \times 60 - 100 = 20$, there were 20 easy problems and $100 - 20 = 80$ hard problems. The difference was $80 - 20 = 60$.

4. (1991-3)

Four girls are starting simultaneously from the same point on a circular track, running at constant speeds which are not necessarily the same. A and B run clockwise while C and D run counterclockwise. A meets C for the first time at the same moment as B meets D for the first time. Prove that A catches up with B for the first time at the same moment as D catches up with C for the first time.

Example.

Four boys are starting simultaneously from the same point on a circular track of length 600 meters. A and B run clockwise while C and D run counterclockwise. They run at different constant speeds, 4 meters per second for A, 3 meters per second for B and 1 meter per second for C. The speed for D is not given. A meets C for the first time at the same moment as B meets D for the first time. How many seconds from the start will it be when D catches up with C for the first time?

Solution:

When A meets C for the first time, they have covered one lap at the combined speed of $4 + 1 = 5$ meters per second. Since B meets D at the same time, the constant speed of D is $5 - 3 = 2$ meters per second. So D pulls 1 meter ahead of C every second. Since the length of the track is 600 meters, it will take D 600 seconds to overtake C for the first time.

Set B : Parity

1. (1990-1)

   Paula numbers the 96 sheets in her notebook page by page in order from 1 to 192. Nick rips out 25 sheets at random and adds together all 50 page numbers. Prove that his sum cannot be equal to 1990.

   Example.

   A faulty money-changer gives five pennies for a nickel and five nickels for a penny. Can someone starts with a single penny and ends up with an equal number of pennies and nickels using only this machine?

   Solution:

   It is impossible. The total number of coins is 1 initially, which is odd. This total increases by 4 each time we use the machine. Hence it remains odd. However, in order to have an equal number of pennies and nickels, the total number of coins must be even.

2. (1982-2)

   A grasshopper jumps 1 cm. Then it jumps 3 cm in the same or the opposite direction. Then it jumps 5 cm in the same or the opposite direction, and so on. Can the grasshopper get back to the starting point on the 25th jump?

   Example.

   A grasshopper jumps 2 cm. Then it jumps 4 cm in the same or the opposite direction. Then it jumps 6 cm in the same or the opposite direction, and so on. Can the grasshopper get back to the starting point on the fifth jump?

   Solution:

   Let the grasshopper start at 0 on the number line and get to 2 after the first jump. The chart below shows where it gets to after the next three jumps.

Second Jump	-2				6			
Third Jump	-8		4		0		12	
Fourth Jump	-16	0	-4	12	-8	8	4	20

   The grasshopper is not at -10 or 10, and cannot return to 0 after the fifth jump.

3. (1984-3)

   On a line containing a segment AB, there are 45 points marked, none of which lie on the segment AB. Prove that the sum of the distances from these points to A is not equal the sum of their distances to B.

Example.

Twelve houses, numbered from 1 to 12, are evenly spaced along one side of a street. Five children live in #1, one child in #2, one in #3, four in #6 and two in #12. The thirteen children meet at one of their houses. Which house should it be if the total distance the children have to travel is as small as possible?

Solution:

Consider the children in pairs from opposite ends of the street. The distance traveled by the first child and the last child is always the distance between #1 and #12, regardless of where the meeting is. Similarly, the distance traveled by the second child and the second last child is also constant, and so on. Since the middle child lives in #3, there is where the meeting should be.

Set C : The Pigeonhole Principle

1. (1985-5)

   Among 300 boots, 100 are of size 8, 100 of size 9 and 100 of size 10. There are 150 right boots and 150 left boots. Prove that one can select at least 50 pairs of boots each consisting of a right boot and a left boot of the same size.

   Example.

   Among 200 boots, 120 are of size 8 and 80 of size 9. There are 100 right boots and 100 left boots. What is the minimum number of pairs of boots each consisting of a right and a left boot of the same size?

   Solution:

   Put away the minimum number of matching pairs of boots and consider the unmatched boots. Those of one size must either all be right boots or all be left boots. If all 80 boots of size 9 are right boots, then we have only 20 pairs of matching boots of size 8. This is indeed minimum since we cannot leave out any more boots of size 9.

2. (1979-3)

   Counters are placed on 25 of the squares of a 6×7 chessboard. Prove that there exists a 2×2 subboard with at least three counters on its four squares.

   Example.

   What is the minimum number of counters that must be placed on the squares of a 3×7 chessboard so that there will be a 2×2 subboard with at least three counters on its four squares.

   Solution:

   If we place 14 counters on the chessboard, one on each square of the top row and the bottom row, there are no 2×2 subboards with the desired property. Suppose we place 15 counters. Divide the chessboard into three 2×2 subboards along with the bottom row and the rightmost column, as shown in Figure 1. Suppose none of the three 2×2 subboards contains at least three counters. Then each must contain two counters. Moreover, the bottom row and the rightmost column must be completely filled. Then the shaded 2×2 subboard contains at least three counters.

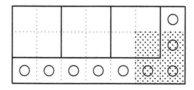

Figure 1

3. (1981-4)

What is the smallest number of counters that must be placed on the squares of a 12×12 chessboard so that the shape in Figure 2 cannot be placed on three unoccupied squares of the chessboard?

Figure 2

Example.

What is the smallest number of counters that must be placed on the squares of a 12×12 chessboard so that no three adjacent squares in the same row or the same column are unoccupied?

Solution:

Since the chessboard may be divided into 48 copies of a 1×3 subboard, we must have at least one counter on each subboard. Figure 3 shows that 48 counters are also sufficient.

Figure 3

Set D : Extremal and Mean Value Principles

1. (1988-2)

   A teacher has a deck of 30 cards numbered from 1 to 30, as does each of 30 students. They all turn over the top cards in their respective decks. Whenever the number on a student's card matches the number on the teacher's card, that student scores 1 point. When all the cards have been turned over, each student scores a different number of points. Prove that one of them scores 30 points.

   Example.

   Various pairs of 30 students shake hands. Prove that there are two students who have shaken hands with the same number of other students.

   Solution:

   The number of students with whom a student can shake hands ranges from 0 to 29 inclusive. Suppose to the contrary that no two of them have shaken hands with the same number of others. Then the respective numbers for them are precisely 0, 1, 2, ..., 29. However, if one of the students have shaken hands with 0 other students, then no student could have shaken hands with 29 others. We have a contradiction.

2. (1982-3)

   The squares of a 5×5 chessboard are painted in one of two colors in an arbitrary way. Prove that there exist 2 rows and 2 columns such that the 4 squares where they intersect are all of the same color.

   Example.

   What is the minimum number of squares of a 3×4 chessboard that must be painted so that there will always be 2 rows and 2 columns such that the 4 squares where they intersect are all painted?

   Solution:

   Figure 4 shows that painting 7 squares are not enough.

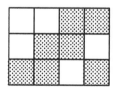

Figure 4

Suppose we paint 8 squares. Suppose all three squares in a column are painted. Then there is at least another column with at least two painted squares, and the desired result is obtained.

Suppose each column contains exactly two painted squares. They may be in any pairs of rows 1, 2 and 3. Since we have three combinations and four columns, the desired result follows from the Pigeonhole Principle.

3. (1984-4)

   Each square of an infinite chessboard is painted in one of 8 colors. Prove that it is possible to place a copy of the shape in Figure 5, possibly rotated or reflected, such that it covers two squares of the same color.

Figure 5

Example.

Each square of an infinite chessboard is painted in one of 3 colors. Prove that it is possible to place a copy of the shape in Figure 6, possibly rotated or reflected, such that it covers two squares of the same color.

Figure 6

Solution:

By the Pigeonhole Principle, two squares in a 2×2 chessboard must have the same color. Any two squares in a 2×2 chessboard can be covered by a copy of the shape in Figure 6, possibly rotated or reflected.

Set E : Inequalities

1. (1989-1)

   Each of the competition for the fifth, sixth, seventh, eighth, ninth and tenth grades consists of seven problems. In each competition, exactly four of the questions do not appear on the competition of any other grade. What is the greatest number of distinct problems that may appear in these six competitions?

   Example.

   Each of the competition for the fifth, sixth, seventh, eighth, ninth and tenth grades consists of seven problems. In each competition, exactly four of the questions do not appear on the competition of any other grade. What is the smallest number of distinct problems that may appear in these six competitions?

   Solution:

   The four problems in each competition which are not duplicated yield a total of $6 \times 4 = 24$ distinct problems. To minimize the number of distinct problems, each of the duplicated problems must appear in all six papers, yielding an additional 3 distinct problems, for a total of $24 + 3 = 27$.

2. (1991-1)

   Each of 40 children in a workshop class has nails, nuts and bolts. There are exactly 15 children with unequal numbers of nails and nuts, and 10 children with equal numbers of nuts and bolts. Prove that there are at least 15 children with unequal numbers of nails and bolts.

   Example.

   Each of 40 children in a workshop class has nails, nuts and bolts. There are exactly 15 children with equal numbers of nails and nuts, and 10 children with equal numbers of nuts and bolts. What is the maximum number of children with equal numbers of nails and bolts?

   Solution:

   Since there are 15 children with equal numbers of nails and nuts, and 10 children with equal numbers of nuts and bolts, at least 5 children must have different numbers of nails and bolts. Hence the maximum number of children with equal numbers of nails and bolts is $40 - 5 = 35$. This may be attained if each of 10 children has a nail, a nut and a bolt, each of 25 children has 1 nail and 1 bolt, while each of the remaining 5 children has 1 nail and 1 nut.

3. (1990-6)

   For any boy in a club, all the girls who know him know one another. Each girl knows more boys than girls, not counting themselves. Prove that there are at least as many boys as girls in this club.

Example.

There are twenty girls in a club and they all know one another. Each girl knows more boys in the club than At least how many boys are in the club?

Solution:

Since each girl knows nineteen other girls and more boys than that, there must be at least twenty boys in the club.

Set F : Many-to-one Correspondences

1. (1979-5)
   A class consists of 31 grade 2 students and some grade 3 students. There are 19 tables at which one or two students may sit. If each boy knows exactly three girls and each girl knows exactly two boys, how many students are there altogether?

   Example.
   Whenever the Walrus ate five oysters, the Carpenter ate three. The total number of oysters they ate was more than one hundred but less than one hundred and ten. Exactly how many oysters did they eat?

   Solution:
   The number of oysters eaten must be a multiple of 5+3=8. The only multiple of 8 between 100 and 110 is 104.

2. (1980-2)
   Each of 23 students is 10, 11, 12 or 13 years old, with at least one of each. Their total age is 253 years. How many 12 year olds are there if there are 1.5 times as many 12 year olds as 13 year olds?

   Example.
   The costs for a pack of chocolate, caramel and toffee are $2, $3 and $4 respectively. Kate spends $20 on these candies. She gets the same number of packs of caramel and of toffee. How many packs of chocolate does she get?

   Solution:
   The total cost of a pack of caramel and a pack of toffee is $7. Since Kate gets the same number of packs of each kind, she must have spent either $7 or $7 × 2 = $14 on them, because she spends less than $21. This amount must be $14 since the cost of a pack of chocolate is an even number. Thus she has an even number of dollars left, and must have bought (20 − 14) ÷ 2 = 3 packs of chocolate.

3. (1991-2)
   A strange rule in a club allows the children only to trade any two marbles for any three others, or any three marbles for any two others. Is it possible to start with 100 red marbles and end up with 100 green marbles, having traded away exactly 1991 marbles in the meantime?

   Example.
   A strange rule in a club allows the children only to trade any two marbles for any three others, or any three marbles for any two others. Starting with 100 red marbles, what is the minimum number of marbles we have to trade away in order to end up with 100 green marbles?

Solution:
Since we start with 100 red marbles and end up with 0 red marbles, all 100 have to be traded away. This minimum can be attained by trading away 60 red marbles for 40 green marbles, and the remaining 40 red marbles for 60 green marbles.

Set G : Arithmetical Problems

1. (1987-4)

   The cost of a hotdog and the cost of a burger are both integral numbers of cents. If each boy buys a hotdog and each girl buys a burger, their total expenditure will be one cent more than if each boy buys a burger and each girl buys a hotdog. There are more boys than girls. How many more?

   Example.

   The cost of a hotdog and the cost of a burger are both integral numbers of cents. If each boy buys a hotdog and each girl buys a burger, their total expenditure will be five cent more than if each boy buys a burger and each girl buys a hotdog. There are more boys than girls. How large can the difference between the numbers of boys and girls be?

   Solution:

   The change in total expenditure is a multiple of the difference between the numbers of boys and girls. Since this change is 5 cent, the number of boys exceeds the number of girls by 1 or 5, and the maximum difference is 5.

2. (1981-1)

   Adam and Betty wrote 54 tests marked out of 3. Checking their records, Adam had as many 3s as Betty had 2s, as many 2s as Betty had 1s, and as many 1s as Betty had 0s. Prove that their averages are not the same.

   Example.

   Adam and Betty wrote 54 tests marked out of 2. Checking their records, Adam had as many 2s are Betty had 1s, and as many 1s as Betty had 0s. Is it possible that their averages are the same?

   Solution:

   It is possible. Each may have 18 of each of 2s, 1s and 0s.

3. (1989-3)

   In a model railway set, there are two kinds of track pieces as shown in Figure 7. They may not be flipped over. When two pieces are put together, the convex head of one must fit into the concave tail of the other. A closed track constructed according to this rule has been taken apart, and one piece has been replaced by a piece of the other kind. Prove that it is now impossible to reassemble the pieces to form a closed track which follows the rule.

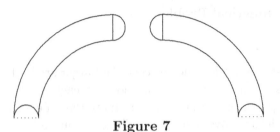

Figure 7

Example.

In a model railway set, there are two kinds of track pieces as shown in Figure 7. They may not be flipped over. When two pieces are put together, the convex head of one must fit into the concave tail of the other. Is it possible to construct a closed track according to this rule

(a) with six pieces of one kind and two pieces of the other kind;

(b) with four pieces of one kind and four pieces of the other kind?

Solution:

(a) Such a construction is shown in Figure 8. The two pieces in the middle are different from the other six pieces.

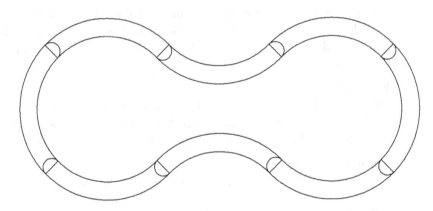

Figure 8

(b) The construction is impossible. The simplest closed track consists of four pieces of one kind. It is impossible to introduce more pieces of the same kind without first introducing some pieces of the other kind. However, every time a piece of the other kind is introduced, we must correct the deviation by introducing a piece of the original kind. It follows that the numbers of pieces of the two kind must differ by exactly 4.

Set H : Divisibility Problems

1. (1986-3)

   Let a and b be positive integers such that $34a = 43b$. Prove that $a + b$ is composite.

   Example.

   Let a and b be positive integers such that $12a = 21b$. Must $a + b$ be composite?

   Solution:

   No, we may have $a = 7$ and $b = 4$, and $a + b = 11$ is prime.

2. (1982-4)

   Prove that if the sum of two positive integers is 770, then their product is not divisible by 770.

   Example.

   Prove that if the sum of two positive integers is 30, then their product is not divisible by 30.

   Solution:

   Suppose to the contrary that the product is divisible by 30. Since the sum is $30 = 2 \times 3 \times 5$, either both numbers are divisible by 2, or neither is. Since the product is divisible by 2, both numbers must be divisible by 2. Similarly, both numbers are divisible by 3, and both are divisible by 5. Hence each is at least 30, and their sum cannot be 30.

3. (1985-4)

   Find 1000 numbers, not necessarily distinct, whose sum is equal to their product.

   Example.

   Find ten numbers, not necessarily distinct, whose sum is equal to their product.

   Solution:

   Factoring the largest number under 10, we have $9 = 3 \times 3$. Then $4 \times 4 - (4 + 4) = (4 - 1) \times (4 - 1) - 1 = 8$. If we take the numbers 4 and 4 and add 8 copies of 1, we will have 10 numbers whose sum and product are both $4 \times 4 = 16$.

4. (1988-4)

   Do there exist two non-zero integers such that one of them is divisible by their sum and the other is divisible by their difference?

   Example.

   Do there exist two non-zero integers such that both of them are divisible by their sum?

29

Solution:
Such a pair of non-zero integers consists of 4 and −2. The sum is 2, which divides both numbers.

Set I : Digital Problems

1. (1982-1)

   A six-digit number is given. How many different seven-digit numbers have the property that if one digit is removed from it, the given six-digit number will be obtained?

   Example.

   How many different five-digit numbers have the property that if one digit is removed from it, the number 1233 will be obtained?

   Solution:

   If the missing digit is from the first place, the number is one of 11233, 21233, ..., 91233. If it is from the second place, the number is one of 10233, 11233, 12233, ..., 19233. If it is from the third place, the number is one of 12033, 12133, 12233, 12333, ..., 12933. If it is from the fourth place, the number is one of 12303, 12313, 12323, 12333, ..., 12393. If it is from the last place, the number is one of 12330, 12331, 12332, 12333, ..., 12339. Note that 11233 and 12233 appear twice while 12333 appear three times. It follows that the total is $9 + 10 \times 4 - (1 + 1 + 2) = 45$.

2. (1984-1)

   Prove that in the 400-digit number 84198419...8419 some digits can be deleted from the beginning and some from the end such that the sum of the remaining digits is equal to 1984.

   Example.

   From the 400-digit number 84198419...8419, delete some digits from the beginning and some from the end such that the sum of the remaining digits is equal to 84.

   Solution:

   The sum of the digits of 8419 is 22. The sum of the digits of the 12-digit number 198419841984 is 66. We need to increase this sum by 18. Thus we delete from the original number the first two digits 84 and keep only the next fifteen digits 198419841984198.

3. (1979-2)

   In 1979, Natasha's age was equal to the sum of the digits of the year when she was born. What year was that?

   Example.

   In 978, Natasha's age was equal to the sum of the digits of the year when she was born. What year was that?

Solution:

The largest digit-sum of any year before 978 is 9+6+9=24, so that Natasha was at most 24 years old. Suppose she was born in year y with digit-sum x. Then $978 - y = x$ so that $x + y = 978$. Note that $x \equiv y \pmod 9$. Hence $2x \equiv 6 \pmod 9$ so that $x \equiv 3 \pmod 9$. Since $x \le 24$, we have $x = 3$, 12 or 21. If $x = 3$, then $y = 978 - 3 = 975$. If $x = 12$, then $y = 978 - 12 = 966$. In both cases, the digit-sum is wrong. If $x = 21$, then $y = 978 - 21 = 957$, and digit-sum is correct. Hence Natasha was born in 1957.

Set J : Tests of Divisibility

1. (1986-6)

   (a) Find a seven-digit number with distinct digits which is divisible by all of its digits.

   (b) Does their exist an eight-digit number with this property?

   Example.
   Can a number be divisible by each of its digits if it consists of

   (a) one 1, one 2, one 3, one 4 and one 5;

   (b) one 1, one 2, one 3 and one 4;

   (c) one 1, one 2 and one 3?

   Solution:

   (a) This is not possible. In order to be divisible by 5, the last digit of this number must be 5. Then it is not divisible by 2.

   (b) This is also not possible because the digit-sum $1 + 2 + 3 + 4 = 10$ is not divisible by 3, so that the number is not divisible by 3.

   (c) This is possible. The number must end in 2, but it can be 132 or 312.

2. (1979-4)
   Each of the digits 0 to 9 is written on a card.

   (a) Prove that from any three of the ten cards, a multiple of 3 with up to three digits can be formed.

   (b) What is the minimum number of cards from which a multiple of 9 with up to nine digits can always be formed?

   Example.
   Each of the digits 0 to 9 is written on a card. What is the minimum number of cards from which a multiple of 4 with up to four digits can always be formed?

   Solution:
   Five cards are not enough because we may draw all five cards with odd digits. Six cards are still not enough because the sixth card may contain 0, 4 or 8. Seven Cards are sufficient. At least one card contains an odd digit. If another card contains 2 or 6, we have a two-digit multiple of 4. Otherwise, we must have at least two of the cards containing 0, 4 and 8, and they form a two-digit multiple of 4.

3. (1983-6)

Four different digits are given. We use each of them exactly once to construct the largest possible four-digit number, and use each of them exactly once to construct the smallest possible four-digit number which does not start with 0. If the sum of these two numbers is 10477, what are the given digits?

Example.

The teacher asks Adam to choose a four-digit number which does not end in 0, obtain another four-digit number by writing the digits of the first number in reverse order. Then the teacher asks Betty to do the same exercise. Adam's answer is 5455 while Betty's answer is 4213. The teacher claims that at least one of them has made a mistake. Which of them must have made a mistake?

Solution:

We claim that a correct answer must be a multiple of 11. Since Adam's answer is not a multiple of 11, he has made a mistake. Betty's answer may come from $1652 + 2561 = 4213$. We now justify the claim. The first digit of the original number is in the thousands place and in the units place in the new number. So its value is contributed $1000 + 1 = 1001$ times to the final sum. The second digit of the original number is in the hundreds place and in the tens place in the new number. So its value is contributed $100 + 10 = 110$ times to the final sum. Similarly, the value of the third digit of the original number is contributed 110 times while the value of the fourth digit of the original number is contributed 1001 times. Since both 1001 and 110 are multiples of 11, the claim is justified.

Set K : Squares and Square Roots

1. (1981-5)

   Does there exist a positive integer such that its square begins with the digits 123456789?

   Example.

   Find a positive integer whose square begins with the digits 123?

   Solution:

   Note that $11^2 = 121$ starts with the digits 12 but not 123. However, $111^2 = 12321$ does.

2. (1985-2)

   A 45-digit number consists of one 1, two 2s, three 3s, ..., and nine 9s. Prove that it is not the square of an integer.

   Example.

   A 10-digit number consists of one 1, two 2s, three 3s and four 4s. Can it be the square of an integer?

   Solution:

   The sum of the ten digits is $1 \times 1 + 2 \times 2 + 3 \times 3 + 4 \times 4 = 30$. This is a multiple of 3 but not a multiple of 9. Hence the ten-digit number is also a multiple of 3 but not a multiple of 9, and cannot be the square of an integer.

3. (1992-3)

   Adam and Betty are of the same age. Adam multiplies his age this year by his age last year. Betty calculates the square of her age next year. Prove that the two answers have different digit-sums.

   Example.

   Adam is two years older than Betty. Can the squares of their ages have the same digit-sum?

   Solution:

   If Adam is 28 and Betty is 26, then the digit-sums of $28^2 = 784$ and $26^2 = 676$ are both 19.

Set L : Cyclic Numbers

1. (1989-5)

   Find two six-digit numbers such that the twelve-digit number obtained when the digits of one are written after the digits of the other is divisible by the product of the original six-digit numbers.

   Example.

   Find two two-digit numbers such that the four-digit number obtained when the digits of one are written after the digits of the other is divisible by the product of the original two-digit numbers.

   Solution:

   Since the two-digit number in front divides itself as well as the four-digit number, it must also divide the two-digit number at the back. The quotient of this division is a single digit such that when it is added to 100, the sum is divisible by the two-digit number at the back. Now 101 is prime, but $102 = 2 \times 3 \times 17$ leads to 1734. Indeed, $\frac{1734}{17 \times 34} = \frac{102}{34} = 3$.

2. (1987-5)

   A six-digit number from 000000 to 999999 is said to be *lucky* if the sum of its first three digits is equal to the sum of its last three digits. What is the minimum length of a block of consecutive numbers which will guarantee the inclusion of at least one lucky number, whatever the first number of the block may be?

   Example.

   Prove that every block of 101 consecutive positive integers contains a number whose first two digits are the same as the last two digits.

   Solution:

   Consider 101, 202, ..., 909, 1010, 1111, 1212, ..., 9999, all of which have the desired property. The difference between any two adjacent ones is always 101. The desired conclusion follows immediately.

3. (1989-2)

   Prove that among the six-digit numbers from 000000 to 999999, there are as many with digit sum 27 as those where the sum of its first three digits is equal to the sum of its last three digits.

   Example.

   Among the four-digit numbers from 0000 to 9999, Adam circles those with digit-sum 18 while Betty circles those where the sum of its first two digits is equal to the sum of its last two digits. Who circles more numbers?

Solution:

The first few numbers Betty circles are 0000, 0101, 0110, 0202, 0211 and 0220. We take each of the last two digits and replace it by the difference when it is subtracted from 9. Hence the corresponding numbers are 0099, 0198, 0189, 0297, 0288 and 0279. All of them have digit-sum 18. This transformation is a one-to-one correspondence between the numbers circled by Adam and those circled by Betty. It follows that neither circles more numbers than the other.

Set M : Problems on Money

1. (1987-2)

   A certain country has only four kinds of bills, worth $1, $10, $100 and $1000. Can exactly half a million bills be worth exactly one million dollars?

   Example.

   Vanya has five bills, each of denomination either $1 or $10. He buys notebooks worth $9 each, as many as he can afford. How much money will he have left?

   Solution:

   If Vanya has only $1 bills, he cannot afford any notebook and has $5 left. Trading each $1 bill for a $10 bill allows him to buy a notebook without changing the amount of money he has left. Hence he will still have $5.

2. (1981-6)

   A certain country has only four kinds of bills, worth $1, $2, $5 and $10. Prove that from a stack of bills totaling $400, an outsider can be paid exactly $300.

   Example.

   A certain country has coins worth any integral number of cents. An *efficient* set of coins is one with which we can pay exactly any integral number of cents up to the total value of the coins. Find all efficient sets of four coins which contains at least one coin worth at least 5 cents.

   Solution:

   We must have at least one 1-cent coin. Suppose we only have one. Then the next one must be a 2-cent coin. Then we do not need a 3-cent coin, and the set is completed with a 4-cent coin and a 8-cent coin. We can reduce (1,2,4,8) to (1,2,4,7), (1,2,4,6), (1,2,4,5), (1,2,3,7), (1,2,3,6), (1,2,3,5), (1,2,2,6), (1,2,2,5), (1,1,3,6), (1,1,3,5), and (1,1,2,5).

3. (1992-6)

   Three counterfeiters print bills of arbitrary integral denominations. Each one prints bills totaling $100, and can pay either of the other two counterfeiters any amount up to $25, perhaps with change. Prove that together they can pay an outsider exactly any amount from $100 to $200.

Example.

Two counterfeiters print bills of arbitrary integral denominations. Each one prints bills totaling $100, and can pay the other any amount up to $25, perhaps with change. Prove that at least one of them has a number of bills with total value between $25 and $50.

Solution:

Suppose one counterfeiter can pay any amount up to $25 to other, perhaps with change. If an amount between $25 and $50 is paid by the first counterfeiter, there is nothing further to prove. If it is between $50 and $75, then the bills of the first counterfeiter not involved in this transaction have total value between $25 and $50. Finally, if the amount paid by the first counterfeiter is between $75 and $100, then the second counterfeiter gives change between $50 and $75. The bills of the second counterfeiter not involved in this transaction have total value between $25 and $50.

Set N : Magic Configurations

1. (1980-1)

   Is it possible to construct a 5 × 6 table with the integers from 1 to 30 such that the sum of the six numbers in each row is constant, and the sum of the five numbers in each column is also constant?

   Example.

   Is it possible to construct a 4 × 6 table with the integers from 1 to 24 such that the sum of the six numbers in each row is constant, and the sum of the four numbers in each column is also constant?

   Solution:

   Such a table is shown in Figure 9.

1	24	5	20	9	16
23	2	19	6	15	10
22	3	18	7	14	11
4	21	8	17	12	13

   Figure 9

2. (1984-2)

   Construct a 4 × 4 table of non-zero numbers such that the sum of the numbers in the four corner squares of any 2 × 2, 3 × 3 or 4 × 4 subtable is 0.

   Example.

   There is a 0 in each of the four corner squares of a 4 × 4 table. Is it possible to fill in the remaining squares with non-zero numbers so that the sum of all numbers along a diagonal of any length is 0?

   Solution:

   Figure 10 shows one of many such tables.

0	−1	−1	0
1	2	2	1
−1	−2	−2	−1
0	1	1	0

   Figure 10

3. (1982-5)

Label the edges of a cube with 1, 2, 3, ..., 12 so that the sum of the labels of the four edges of each of the six faces is the same.

Example.

Label the vertices of a cube with 1, 2, 3, ..., 8 so that the sum of the labels of the four vertices of each of the six faces is the same.

Solution:

Figure 11 shows one of many possible labelings.

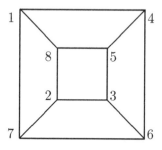

Figure 11

Set O : Logic Problems

1. (1983-3)
   Each of Benny, Denny, Kenny and Lenny either always lies or always tells the truth. Benny claims that Denny is a liar. Lenny claims that Benny is a liar. Kenny claims that both Benny and Denny are liars. Kenny also claims that Lenny is a liar. Which of them always lies and which of them always tells the truth?

   Example.
   Each of Benny and Denny either always lies or always tells the truth. Is it logically possible for Benny and Denny to claim that the other is a liar?

   Solution:
   It is logically possible if exactly one of them is a liar. We would not know which one is the liar.

2. (1991-4)
   Baron Münchhausen hunts ducks everyday. One day, he declares, "Today, I will bring home more ducks than two days ago but fewer than one week ago." For at most how many consecutive days can the Baron say this without telling a lie?

   Example.
   The row of numbers -5, 7, -5, -5, 7, -5, 7, -5, -5, 7, -5 have the property that the sum of any five adjacent numbers is negative and the sum of any eight adjacent numbers is positive. Does there exist a longer row of numbers with this property, where -5 and 7 may be replaced by two other numbers?

   Solution:
   Suppose such a row of twelve numbers exist. Construct a 5×8 table as follows. Use the first eight numbers to form the first row. Omit the first number and use the next eight numbers to form the second row. Omit the first two numbers and use the next eight numbers to form the third row. Omit the first three numbers and use the next eight numbers to form the fourth row. Omit the first four numbers and use the last eight numbers to form the fifth and last row. Now the sum of all the numbers in each row is positive but the sum of all the numbers in each column is negative. This is impossible.

3. (1983-5)

Baron Münchhausen has a time machine which allows a jump from March 1st to November 1st of any other year, from April 1st to December 1st, from May 1st to January 1st and so on. He cannot arrive and then depart on the same day. The Baron starts and ends his time travel on April 1st, and claims that he has been away for 26 months. Prove that he is mistaken.

Example.

Sandy got on the exercise bike at noon. She always rode for 18 minutes in a stretch, and then rested a fixed positive integral number of minutes before getting back on. She got off the last time at two in the afternoon. Could the total number of minutes during which she had rested be 30?

Solution:

Suppose this was possible. Then she had ridden $120 - 30 = 90$ minutes. Since each stretch of riding lasts 18 minutes, she had rested 4 times. However, 30 is not a multiple of 4.

Set P : Coin Weighing Problems

1. (1985-1)

   Each of 68 coins has a different weight. In 100 weighings on a standard balance, find the heaviest and lightest coins.

   Example.

   Each of 13 coins have a different weight. What is the minimum number of weighings needed to find the heaviest coin?

   Solution:

   In order to find the heaviest coin, we must eliminate 12 coins. Since each weighing eliminates exactly one coin, 12 weighings are both necessary and sufficient.

2. (1989-4)

   Each of 32 coins has a different weight. In 35 weighings on a standard balance, find the heaviest and the second heaviest coins.

   Example.

   We have four coins with different weights. Give a complete ranking in descending order of weight in five weighings.

   Solution:

   In the first two weighings, we weigh the coins in pairs. In the third and fourth weighing, we weigh the heavier coins in the first two weighings against each other, and the lighter coins in the first two weighings against each other. We will now know which are the heaviest and the lightest coins. A fifth weighing between the other two decides which coin is the second heaviest and which coin is the second lightest.

3. (1990-2)

   Among 101 coins, 100 are genuine and have the same weight. The weight of the counterfeit coin is different from that of a genuine coin. Determine whether the counterfeit coin is heavier or lighter in two weighings on a standard balance. The identification of the counterfeit coin is not required.

   Example.

   Among four coins, three are genuine and have the same weight. The weight of the counterfeit coin is different from that of a genuine coin. Determine whether the counterfeit coin is heavier or lighter in two weighings on a standard balance. The identification of the counterfeit coin is not required.

   Solution:

   Weigh two coins against the other two. This cannot result in equilibrium. Weigh the two coins on the heavier side. If there is equilibrium, the counterfeit coin is light. Otherwise, it is heavy.

4. (1980-4)

Seven genuine coins have the same weight. Two counterfeit coins also have the same weight. A counterfeit coin is heavier than a genuine coin. Identify the fake coins using a standard balance at most four times.

Example.

A counterfeit coin is heavier than a genuine coin. We have three coins, and we are told the number of counterfeit coins among them. What is the minimum number of weighings necessary to identify the counterfeit coins?

Solution:

If this number is zero or three, no weighings are required. Suppose it is one. We weigh one coin against another. If we have equilibrium, the third one is the counterfeit coin. Otherwise, the heavier one is the counterfeit coin. Finally, suppose the number of counterfeit coins is two. Again, we weigh one coin against another. If we have equilibrium, both coins are counterfeit. Otherwise, the heavier one along with the third one are the counterfeit coins.

Set Q : Geometric Configurations

1. (1986-4)
   Find a configuration of several identical round coins on a table so that each coin touches exactly three other coins.

   Example.
   Find a configuration of sixteen identical round coins on a table so that each of ten coins touches exactly three other coins, each of three coins touches exactly four other coins while each of the remaining three coins touches exactly two other coins.

   Solution:
   Figure 12 shows such a configuration.

   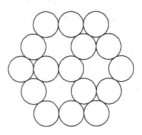

   Figure 12

2. (1986-2)
   There are 44 queens on a standard chessboard. Prove that each queen attacks at least one other queen.

   Example.
   What is the maximum number of bishops on a standard chessboard such that no two attack each other?

   Solution:
   We can place a bishop on each square of the bottom row and on each non-corner square of the top row. No two of these 14 bishops can attack each other. Consider the 15 diagonals running from northwest to southeast. Each can contain at most one bishop. Moreover, the two diagonals of length 1 cannot both contain a bishop. So 14 is indeed the maximum.

3. (1992-2)
   In a heptagonal castle, each of the seven sides is a straight wall and there is a watchtower at each of the seven vertices. The guards stay in the watchtowers. Each guard watches over both walls meeting at that watchtower. What is the minimum number of guards required so that each wall is watched over by at least 7 guards?

Example.

In a square castle, each of the four sides is a straight wall. There is a watchtower at each of the four vertices and at the midpoint of each of the four sides. The guards stay in the watchtowers. Each guard watches either the wall containing that tower, or over both walls meeting at that watchtower. What are the possible number of guards if each wall is watched over by exactly 9 guards?

Solution:

No guard can watch over both the north wall and the south wall. Since 9 guards are needed to watch over each, the minimum number of guards is 18. This can be accomplished as shown in Figure 13 on the left. On the other hand, since each wall is watched over by 9 guards, the maximum number of guards is 36. This can be accomplished as shown in Figure 13 on the right. All numbers between 18 and 36 are possible. To get a total of 19, replace 1 guard at a corner watchtower by 2 guards, 1 in each adjacent side tower. This is shown in Figure 13 in the middle. We can continue to replace guards until their total number reaches 36.

Figure 13

Set R : Problems on Coloring

1. (1980-3)

   On the line AB, 200 points are chosen symmetrically with respect to the midpoint of AB. Half of the points are red, and half are blue. Prove that the sum of the distances from A to all the red points is equal to the sum of the distances from B to all the blue points.

   Example.

   Thirteen houses, numbered from 1 to 13, are evenly spaced along one side of a street. Two children live in #2, one child in #3, one in #6, one in #8, one in #11 and two in #12. Among the ten children, five are girls and five are boys. The girls meet at #1 and the boys meet at #13. Prove that the total distance the boys have to travel is always equal to the total distance the girls have to travel?

   Solution:

   Consider the children in pairs from opposite ends of the street. Suppose one of the pair is a girl and the other is a boy. Then the distances traveled by these two children are equal. Suppose both are boys or both are girls. Then the total distance the pair has traveled is equal to the distance between #1 and #13. Since the number of girls is equal to the number of boys, the number of boy-pairs must be equal to the number of girl-pairs. The desired conclusion follows.

2. (1981-2)

   Does there exist a number consisting of each of the digits 1 to 9 exactly once such that between any two digits differing by 1, there are an odd number of other digits?

   Example.

   Can we place two copies of each of 1, 2, 3 and 4 in a row, with 1 other digit between the two 1s, 2 other digits between the two 2s, 3 other digits between the two 3s, and 4 other digits between the two 4s?

   Solution:

   We first put down the two 4s with 4 spaces in between. It is easy to discover that exactly one of the two 3s is between the two 4s. From here, it is not hard to complete the construction of the row: 2 3 4 2 1 3 1 4.

3. (1992-5)

   A circle is divided into 27 equal arcs by 27 points. Each point is either white or black. No two black points are adjacent or separated by only one white point. Prove that three of the white points are the vertices of an equilateral triangle.

Example.

A circle is divided into 9 equal arcs by 9 points. Each point is either white or black. No two black points are adjacent or separated by only one white point. Prove that three of the white points are the vertices of an equilateral triangle.

Solution:

There are three sets of points that are the vertices of an equilateral triangle. If None of them consists of three white points, then there are at least 3 black points and at most 6 white points. Since there are at least two white points between two black points, there are exactly 3 black points, with exactly two white points between two of them. Now the black points form one of the three sets. Thus any of the other two sets consists of three white points which are the vertices of an equilateral triangle.

Set S : Tournament Problems

1. (1983-1)

   In a chess tournament, each participant plays every other participants exactly once. Each participant gets 1 point for a win, $\frac{1}{2}$ point for a draw and 0 points for a loss. At most how many of the 30 participants can score 18 points or more?

   Example.

   In a chess tournament, each of eight participants plays a game against each of the others. A win is worth 1 point, a draw $\frac{1}{2}$ point, and a loss 0 points. At the end of the tournament, each participant has a different total score, and that of the participant in second place is equal to the sum of those of the bottom four participants. What is the result of the game between the participants in third and seventh places?

   Solution:

   The bottom four participants played six games among themselves, and therefore the sum of their scores was no less than 6 points even if they lost every game against the top four. Hence the score of the runner-up was at least 6 points. If the winner scored 7 points, then the runner-up lost the game against the winner, and finished with no more than 6 points. If the winner scored 6.5 points, again the runner-up could have no more than 6 points. Hence the runner-up had exactly 6 points, and the bottom four participants had exactly 6 points among themselves, meaning that they lost every game against the top four participants. In particular, the game between the participants in third and seventh places was won by the player in the third place.

2. (1992-1)

   In a tournament, each participant plays every other participants exactly once. Each participant gets 1 point for a win, 0 points for a draw, and -1 point for a loss. One of the participants finishes the tournament with 7 points and another with 20. Prove that there is at least one drawn game.

   Example.

   In a tournament, each of six participant plays every other participants exactly once. Each participant gets 1 point for a win, 0 points for a draw, and -1 point for a loss. Construct such a tournament in which one of the participants finishes the tournament with 1 point and another with 4.

Solution:

A possible tournament is shown in the chart below.

| Tournament | Opponents | | | | | | Total |
Record	A	B	C	D	E	F	Score
Participant A	–	1	1	1	1	0	4
Participant B	−1	–	−1	1	1	1	1
Participant C	−1	1	–	1	1	1	3
Participant D	−1	−1	−1	–	1	1	−1
Participant E	−1	−1	−1	−1	–	−1	−5
Participant F	0	−1	−1	−1	1	–	−2

3. (1991-6)

In a tournament without draws, every two of the nine teams play against each other exactly once. Must there always be two teams such that every other team has lost to either or both of them?

Example.

In a tournament without draws, every two of the teams play against each other exactly once. Prove that the participants may be lined up so that each beats the next one in line.

Solution:

We build this line from scratch. Start with any two participants. Since there are no draws, one of the must win and becomes the head of the line. Now add a new participant. If she beats the participant at the head, she becomes the new head. If not, she checks if she beats the next participant in line. If she does, she can be inserted into the line before that participant. If she does not beat anyone in the line, she becomes the new tail of the line. Adding one participant at a time, we can build a complete line for the tournament.

1. (1992-4)

Fyodor collects coins. No coin in his collection is more than 10 cm in diameter. He keeps all the coins arranged side by side in a rectangular box of size 30 cm by 70 cm. Prove that he can fit all of his coins in another rectangular box of size 40 cm by 60 cm.

Example.

Fyodor collects coins. No coin in his collection is more than 10 cm in diameter. He keeps all the coins arranged side by side in a rectangular box of size 10 cm by 40 cm. He acquires a new coin of diameter 25 cm and a second one of diameter 10 cm. Prove that he can fit all of his coins in a square box of size 35 cm by 35 cm.

Solution:

Divide the square box into four sections. Put the bigger new coin in the northeast section which is 25 cm by 25 cm. Put the smaller new coin in the southwest section which is 10 cm by 10 cm. Divide the old box into a left section which is 10 cm by 15 cm, a middle section which is 10 cm by 10 cm and a right section which is 10 cm by 15 cm. Since none of the old coins has diameter more than 10 cm, no coin is in both the left section and the right section. Now transfer all the old coins entirely within the left and middle sections to the southeast section of the square box which is 10 cm by 25 cm, and all the other old coins to the northwest section of the square box which is also 10 cm by 25 cm.

2. (1979-1)

Pack fifteen 2×3 chocolate pieces into a 7×13 box, leaving a 1×1 hole.

Example.

What is the maximum number of pieces of the of the shape in Figure 14 that can be packed into a 6×6 box?

Figure 14

Solution:

Since the area of the box is 36 while the area of the shape is 4, at most 9 pieces can possibly fit. However, since the corner squares cannot be used, at most 8 pieces can be packed, as shown in Figure 15.

Figure 15

3. (1990-3)
   Is it possible to pack thirty-nine 5×11 chocolate pieces into a 39×55 box?

Example
How many 3×4 chocolate pieces can we pack into an 11×11 box, if the sides of the pieces must be parallel or perpendicular to the sides of the box?

Solution:
Since $11 \times 11 = 121$, $3 \times 4 = 12$ and $121 = 10 \times 12 + 1$, it would appear that we can pack as many as ten pieces into the box. Experimentation yields a nine-piece packing as shown in Figure 16 on the left. We claim that this is best possible. Consider the nine shaded squares in Figure 16 on the right. No matter where a piece is placed, it must cover at least one shaded square. Since we only have 9 shaded squares, we cannot place 10 pieces.

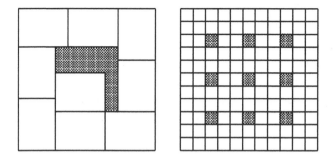

Figure 16

Set U : Dissection Problems

1. (1980-5)

   Dissect a square into convex pentagons.

   Example.

   Is it possible to dissect an equilateral triangle into a triangle, a convex quadrilateral, a convex pentagon and a convex hexagon?

   Solution:

   Each of the four pieces must have at most three sides inside the equilateral triangle, since such sides must also belong to other convex polygons. This means that the hexagon has three sides along the perimeter, the pentagon has two sides and the quadrilateral has one. A possible dissection is shown in Figure 17.

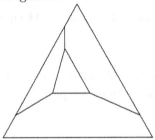

   Figure 17

2. (1981-3)

   Place nine points inside a triangle so that there are twelve points altogether. In addition to the three sides of the triangle, join some other pairs of these twelve points with non-intersecting segments so that each point is joined to five others, and the original triangle is divided into triangles.

   Example.

   Place three points inside a triangle so that there are six points altogether. In addition to the three sides of the triangle, join some other pairs of these six points with non-intersecting segments so that each point is joined to four others, and the original triangle is divided into triangles.

Solution:
The configuration may be the net of an octahedron, as shown in Figure 18.

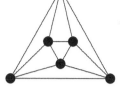

Figure 18

3. (1983-2)

Ten 10×20 chocolate pieces are cut up into twenty triangles. Pack them into a square box, leaving no empty space.

Example.
Eight 10×10 chocolate pieces are cut up into sixteen triangles. Pack them into a square box, leaving no empty space.

Solution:
The packing can be done as shown in Figure 19.

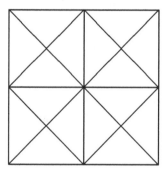

Figure 19

Set V : Graph Theory

1. (1985-3)

   A traveler departed from his home city A to the city B farthest from A. From B, he departed to the city C farthest from B, and so on. Prove that if C and A are different cities, then the traveler will never reach home.

   Example.

   The pairwise distances of nine cities are distinct. From each city, a traveler departs to visit the nearest city. Prove that some city is visited by at least two travelers.

   Solution:

   Consider the two cities at the smallest distance apart. Their travelers visit the other city in this pair. If one of them is visited by a third traveler, we have the desired conclusion. Suppose they are far enough apart from the other cities that no other traveler comes their way. We may discard them and work on the remaining seven cities. We started with an odd number of cities. Discarding two cities at a time, their number remains odd. When we are down to the last city, the traveler from there must visit some city already visited by another traveler.

2. (1980-6)

   The map of a subway system is a convex polygon in which no three diagonals are concurrent. There is a station at each vertex and at every intersection of two diagonals. Train runs along entire diagonals, but not necessarily every diagonal. If each station lies on the route of at least one train, prove that it is possible to go from any station to any other station, changing trains at most twice.

   Example.

   Figure 20 shows the map of a subway system with 21 stations and 9 lines. The three shown as dotted lines are closed for repairs. Find all pairs of stations such that it is not possible to go from one to the other, changing trains at most once.

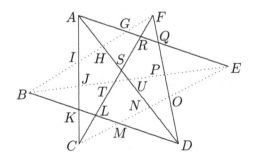

Figure 20

Solution:

Routine checking shows that such pairs are (B, E), (B, G), (E, M), (G, M), (I, O), (I, P), (J, O) and (J, P).

3. (1987-3)

Each of six cities is connected to the others by five roads. Show that it is possible for the roads to intersect only three times with exactly two roads crossing over at each intersection. Junctions at the cities are not considered intersections.

Example.

Each of three cities is connected to each of three towns by a road. Show that it is possible for the nine roads to intersect only once with exactly two roads crossing over at the intersection. Junctions at the cities or towns are not considered intersections.

Solution:

A possible arrangement is shown in Figure 21. The cities are marked with black dots.

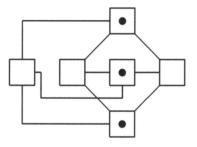

Figure 21

Set W : Numerical Solitaire Games

1. (1984-5)

   A solitaire starts with a circle divided into six sectors each containing a number as shown in Figure 22. In each move, one may add 1 to the numbers in any two adjacent sectors. Prove that it is impossible to make all six numbers equal.

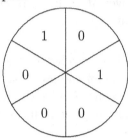

Figure 22

Example.

A solitaire starts with a circle divided into six sectors each containing a number as shown in Figure 23. In each move, one may add 1 to the numbers in any two adjacent sectors. Is it possible to make all six numbers equal?

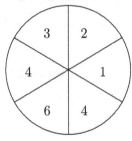

Figure 23

Solution:

This is possible. Increase the adjacent pair (4,1) to (6,3) and the adjacent pair (3,4) to (5,6). Now the numbers are 6, 3, 2, 5, 6 and 6 in cyclic order. Increase the adjacent pair (3,2) to (6,5), and finally increase the adjacent pair (5,5) to (6,6).

2. (1988-1)

   A solitaire game starts with a 0 in each square of a 3×3 table. In each move, one may add 1 to all numbers in any of the four 2×2 subtables. Is it possible to obtain the table in Figure 24?

4	9	5
10	18	12
6	13	7

Figure 24

Example.

A solitaire game starts with a 0 in each square of a 3×3 table. In each move, one may add 1 to all numbers in any of the four 2×2 subtables. Is it possible to obtain the table in Figure 25?

2	5	3
6	14	8
4	11	5

Figure 25

Solution:

This is impossible as the number between 4 and 5 must be $4 + 5 = 9$ and not 11.

3. (1987-1)

A solitaire game starts with the 4×4 table in Figure 26 on the left. In each move, one may add 1 to all the numbers in any row or subtract 1 from all the numbers in any column. How can one obtain the table in Figure 26 on the right?

1	2	3	4
5	6	7	8
9	10	11	12
13	14	15	16

1	5	9	13
2	6	10	14
3	7	11	15
4	8	12	16

Figure 26

Example.

A solitaire game starts with the 3×3 table in Figure 27 on the left. In each move, one may add 1 to all the numbers in any column or subtract 1 from all the numbers in any row. How can one obtain the table in Figure 27 on the right?

1	2	3
4	5	6
7	8	9

1	4	7
2	5	8
3	6	9

Figure 27

Solution:

Add 4 to every number in the first column, 6 to every number in the second column and 8 to every number in the third column. Then subtract 4 from every number in the first row, 6 from every number in the second row and 8 from every number in the third row.

4. (1983-4)

A solitaire game starts with eight numbers arranged in a circle. Each is either 1 or -1, and the choice is arbitrary. In each move, one can multiply any blocks of adjacent numbers of length 3 by -1. Prove that one can make all eight numbers equal to 1.

Example.

A solitaire game starts with nine numbers arranged in a circle. Each is either 1 or -1, and the choice is arbitrary. In each move, one can multiply any blocks of adjacent numbers of length 3 by -1. Is it always possible to make all nine numbers equal to 1?

Solution:

This is not always possible. Suppose we start with 1, 1, -1, 1, 1, -1, 1, 1 and -1. The product of the first, the fourth and the seventh numbers is 1. The product of the second, the fifth and the eighth numbers is also 1. However, the product of the third, the sixth and the ninth numbers is -1. In each move, all these three products change signs. Hence they can never be all the same. It follows that the nine numbers cannot all be the same.

Set X : Geometric Solitaire Games

1. (1986-1)

A solitaire game starts with the cards numbered 7, 8, 9, 4, 5, 6, 1, 2 and 3 placed in a row in that order. In each move, one may remove a block of any numbers of adjacent cards, reverse their order and put them back in the row. Show a sequence of moves which puts the cards in the order 1, 2, 3, 4, 5, 6, 7, 8 and 9.

Example.

A solitaire game starts with the cards numbered 7, 4, 1, 8, 5, 2, 9, 6 and 3 placed in a row in that order. In each move, one may remove a block of any numbers of adjacent cards, reverse their order and put them back in the row, not necessarily at the same position. How can one order the cards 1, 2, 3, 4, 5, 6, 7, 8 and 9?

Solution:

This can be accomplished in four moves as shown in Figure 28. The moving cards are shaded.

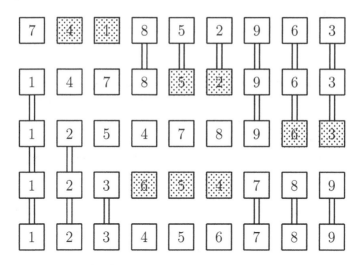

Figure 28

2. (1988-5)

A solitaire game starts with a pile of 1001 counters. In each move, a pile with at least three counters is chosen. One counter is removed while the remaining counters are split into two piles, not necessarily equal in size. Is it possible that after a number of moves, each remaining pile contains exactly three counters?

Example.

A solitaire game starts with a pile of 11 counters. In each move, a pile with at least three counters is chosen. One counter is removed while the remaining counters are split into two piles, not necessarily equal in size. Is it possible that after a number of moves, each remaining pile contains exactly three counters?

Solution:

In the first move, remove one counter from the pile of 11 and split the remaining counters into piles 3 and 7. In the second move, remove one counter from the pile of 7 and split the remaining counters into piles of 3 and 3. The task is accomplished.

3. (1985-6)

A solitaire game starts with ten boxes numbered from 1 to 10 arranged randomly in two stacks. In each move, one may take several boxes from the top of a stack and put them on top of the other stack, which may be empty. Prove that one can form a stack with the ten boxes ordered from 1 at bottom to 10 at the top in 19 moves.

Example.

A solitaire game starts with three boxes numbered 1, 3 and 5 from bottom to top in one stack, and three boxes numbered 2, 4 and 6 from bottom to top in another stack. In each move, one may take several boxes from the top of a stack and put them on top of the other stack, which may be empty. Form a stack with the six boxes ordered from 1 at bottom to 6 at the top in nine moves.

Solution:

In the first move, put 5 and 3 on top of the second stack. In the second move, put the second stack on top of the first stack. In the third move, put 5, 3, 6 and 4 on top of the empty second stack. In the fourth move, put 5 and 3 on top of the first stack. In the fifth move, put 5 on top of the second stack. In the sixth move, put the second stack on top of the first stack. In the seventh move, put 5 and 6 on top of the empty second stack. In the eighth move, put 5 on top of the first stack. In the ninth move, put 6 on top of the first stack.

4. (1988-6)

A solitaire game starts with an 8×8 chessboard in which all 64 squares are white. In each move, one may enter the chessboard on a border square, visit parts of the chessboard going between two squares with a common side, and exit via a border square. The color of a square is changed from white or black or vice versa every time it is visited. Is it possible to create the usual alternating color pattern on the chessboard?

Example.

A solitaire game starts with a 3 × 3 chessboard in which all 9 squares are white. In each move, one may enter the chessboard on a border square, visit parts of the chessboard going between two squares with a common side, and exit via a border square. The color of a square is changed from white or black or vice versa every time it is visited. Is it possible to make only the central square black?

Solution:

Enter the chessboard on a square adjacent to the central square, making it black. Move on to the central square, making it black. Return to the entry square and make it white again, and then exit the chessboard.

Set Y : Numerical Two-Player Games

1. (1982-6)

   Anna and Boris play a game with a stack of 100 counters. Anna goes
   first, and moves alternate thereafter. In each move, a player divides
   a stack of at least two counters into two smaller piles. The loser is
   the player without a move, when each stack consists of exactly one
   counter. Prove that Anna must win this game.

   Example.

   Anna and Boris play a game with a 1×3, a 1×5 and a 1×7 chessboards.
   A counter is placed on the leftmost square of each chessboard. Anna
   goes first, and moves alternate thereafter. In each move, a player
   advances the counter on one of the chessboard one square to the right.
   No further advancement is possible if the counter has reached the
   rightmost square. The player without a move loses the game. Prove
   that Boris must win this game.

   Solution:

   There are 12 legal moves, regardless of who makes them. Since Boris
   goes second, he will make the last move.

2. (1990-4)

   Anna and Boris play a game starting with the number 1234. Anna
   goes first, and turns alternate thereafter. In each turn, the player
   subtracts from the number one of its non-zero digits. A player wins
   if the number is reduced to 0. Who has a winning strategy, Anna or
   Boris?

   Example.

   Anna and Boris play a game starting with the number 100. Anna
   goes first, and turns alternate thereafter. In each move, a player may
   subtract 1, 2 or 3 from the number. A player wins if the number is
   reduced to 0. Which player has a winning strategy, Anna or Boris?

   Solution:

   It is good to leave behind the number 0 since this is the winning
   position. It is not good to leave behind the number 1, 2 or 3 because
   the opponent can then win. It is good to leave behind 4 since the
   opponent cannot win, but must leave behind a number from which you
   can win. The same argument shows that it is good to leave behind
   multiples of 4. Since 100 is already a multiple of 4, Boris has a winning
   strategy, subtracting 3, 2 and 1 whenever Anna subtracts 1, 2 or 3
   respectively.

3. (1984-6)

Anna and Boris play a game with the numbers from 1 to 100 written in order in a row. Anna goes first, and turns alternate thereafter. In each move, a player puts one of the operation signs $+$, $-$ and $\times$ between any two numbers which do not already have an operation sign in between them. After 99 operation signs have been placed, the value of the expression is computed. Anna wins if this value is odd, and Boris wins if it is even. Prove that Anna has a winning strategy.

Example.

Anna and Boris play a game with the numbers from 1 to 6 written in order in a row. Anna goes first, and turns alternate thereafter. In each move, a player puts one of the operation signs $+$ and $\times$ between any two numbers which do not already have an operation sign in between them. After five operation signs have been placed, the value of the expression is computed. Anna wins if this value is odd, and Boris wins if it is even. Prove that Anna has a winning strategy.

Solution:

Anna starts by adding 5 and 6, resulting in the numbers (1 2 3 4 11). This is an alternating sequence which begins and ends in odd numbers. Suppose that Boris inserts a $\times$ sign, resulting in (2 3 4 11), (1 6 4 11), (1 2 12 11) or (1 2 3 44). Anna converts them to (5 4 11), (1 10 11), (1 14 11) or (1 2 47) by inserting a $+$ sign. Suppose that Boris inserts a $+$ sign, resulting in (3 3 4 11), (1 5 4 11), (1 2 7 11) or (1 2 3 15). Anna converts them to (9 4 11), (5 4 11), (1 2 77) or (1 2 45) by inserting a $\times$ sign. In all cases, she again has an alternating sequence which begins and ends in odd numbers, and shortened by two digits. She wins in the next move by inserting the sign different from the one Boris inserts.

Set Z : Geometric Two-Player Games

1. (1987-6)

   Anna and Boris play a game on a 9×9 chessboard. Anna goes first and turns alternate thereafter. In each move, Anna puts a red counter on a vacant square while Boris puts a blue counter on a vacant square. When the board is completely filled, a row with more red counters than blue counters is called a red row, and a blue row otherwise. Red and blue columns are similarly defined. The score for Anna is the sum of the numbers of red rows and red columns while that for Boris is the sum of the numbers of blue rows and blue columns. What is the highest possible score for Anna?

   Example.

   Anna and Boris play a game on a 3×3 chessboard. Anna goes first and turns alternate thereafter. In each move, Anna puts a red counter on a vacant square while Boris puts a blue counter on a vacant square. When the board is completely filled, a row with more red counters than blue counters is called a red row, and a blue row otherwise. Red and blue columns are similarly defined. The score for Anna is the sum of the numbers of red rows and red columns while that for Boris is the sum of the numbers of blue rows and blue columns. What is the highest possible score for Anna?

   Solutions:

   We first show that Anna can score as high as 4 points. She starts by putting a red counter in the central square. Thereafter, she puts a red counter in the square symmetric, with respect to the central square, to the square where Boris has just put a blue counter. She will score row 2 and column 2, one of row 1 and 3 and one of columns 1 and 3. We now show that Boris can hold Anna down to 4 points. Whenever she puts a red counter other than in the central square, he puts a blue counter in the square symmetric to the square where Anna has just put a red counter. If Anna puts a red counter in the central square, Boris makes an arbitrary move. If Anna she puts a red counter in the square symmetric to the square where Boris has just put a blue counter, Boris makes an arbitrary move but not in the central square. Eventually, Anna must put a red counter in the central square. The scoring is exactly the same as before.

2. (1989-6)

   Anna and Boris play a game on a 10×10 chessboard. Anna goes first, and turns alternate thereafter. In each move, a player puts either a red counter or a green counter on a vacant square. A player wins by completing a block of three adjacent counters of the same color along a row, a column or a diagonal. Which player, if either, has a winning strategy?

   Example.

   Anna and Boris play a game on a 3×3 chessboard, with Anna going first. In each move, Anna puts a red counter on a vacant square while Boris puts a blue counter on a vacant square. The player whose counters occupy four squares in the pattern (possibly rotated or reflected) shown in Figure 29 wins the game. Who has a winning strategy?

   Figure 29

   Solution:

   It is clear that the game cannot be won without taking the central square. Hence only Anna can win, and she must start with taking that square. Now Boris can stop her if he can take both squares on either side of the central square. This determines the winning strategy for Anna. After taking the central square, just take the square symmetry, with respect to the central square, to the one Boris just took.

3. (1991-5)

   Anna and Boris play a game with a red stick, a white stick and a blue stick, each of which is 1 meter long. Anna starts by breaking the red stick into three pieces. Then Boris breaks the white stick into three pieces. Finally, Anna breaks the blue stick into three pieces. She wins if she can use the nine pieces to form three triangles with sides of different colors. Can Boris stop her from winning?

   Example.

   Anna and Boris play a game with two sticks each of which is 1 meter long. Anna starts by breaking one stick into two pieces. Then Boris breaks the other stick into two pieces. He wins if he can use three of the four pieces to form a triangle. Can Anna stop him from winning?

Solution:

Anna cannot stop Boris from winning if he breaks his sticks into pieces of lengths $\frac{1}{2}$ meter. No matter how Anna breaks her stick, each piece will be shorter than 1 meter and can be combined with the two pieces of Boris to form a triangle.

Chapter Three: Carry Out the Plan

1979

1. Pack fifteen 2×3 chocolate pieces into a 7×13 box, leaving a 1×1 hole.

   Solution:
   One way is shown in Figure 1.

   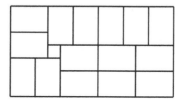

   Figure 1

2. In 1979, Natasha's age was equal to the sum of the digits of the year when she was born. What year was that?

   Solution:
   The largest digit-sum of any year before 1979 is $9+9+9=27=1+8+9+9$, so that Natasha was at most 27 years old. Suppose she was born in year y with digit-sum x. Then $1979 - y = x$ so that $x + y = 1979$. Note that $x \equiv y \pmod 9$. Hence $2x \equiv 8 \pmod 9$ so that $x \equiv 4 \pmod 9$. Since $x \leq 27$, we have $x = 4$, 13 or 22. If $x = 4$, then $y = 1979 - 4 = 1975$. If $x = 13$, then $y = 1979 - 13 = 1966$. In both cases, the digit-sum is wrong. If $x = 22$, then $y = 1979 - 22 = 1957$, and digit-sum is correct. Hence Natasha was born in 1957.

3. Counters are placed on 25 of the squares of a 6×7 chessboard. Prove that there exists a 2×2 subboard with at least three counters on its four squares.

   Solution:
   Suppose no such subboard exists. Divide the chessboard into the top row of 6 squares and nine 2×2 subboard underneath. Now each subboard contains at most 2 counters. Even if there are 6 counters in the top row, the total comes only to 24. This is a contradiction.

4. Each of the digits 0 to 9 is written on a card.

   (a) Prove that from any three of the ten cards, a multiple of 3 with up to three digits can be formed.

69

K. Garaschuk, A. Liu, *Grade Five Competition from the Leningrad Mathematical Olympiad*, Problem Books in Mathematics, https://doi.org/10.1007/978-3-030-52946-8_3

(b) What is the minimum number of cards from which a multiple of 9 with up to nine digits can always be formed?

Solution:

(a) If any of 0, 3, 6 or 9 is on one of the three cards, we have a one-digit multiple of 3. Suppose this is not the case. If we have all of 1, 4 and 7 or all of 2, 5 and 8, we have a three-digit multiple of 3 since 1+4+7=12 and 2+5+8=15 are multiples of 3. If this is also not the case, we take one number from each triple. Their sum will be a multiple of 3, and they will form a two-digit multiple of 3.

(b) Four cards may not be enough as the numbers may be 1, 3, 4 and 7. We claim that five cards are enough. If either 0 or 9 is on one of them, we have a one-digit multiple of 9. Henceforth we assume that this is not the case. Consider the four pairs of numbers (1,8), (2,7), (3,6) and (4,5). If we take five cards, the Pigeonhole Principle guarantees that we will have both numbers of one of those four pairs. They will form a two-digit multiple of 9, justifying our claim.

5. A class consists of 31 grade 2 students and some grade 3 students. There are 19 tables at which one or two students may sit. If each boy knows exactly three girls and each girl knows exactly two boys, how many students are there altogether?

Solution:
The ratio of boys to girls is 2:3. Hence the total number of students is a multiple of 5. It is greater than 31 and less than $19 \times 2 = 38$. Hence it must be 35.

1. Is it possible to construct a 5×6 table with the integers from 1 to 30 such that the sum of the six numbers in each row is constant, and the sum of the five numbers in each column is also constant?

   Solution:
   There are 15 odd numbers among the 30, so that the total is an odd number. This is not divisible by 6, and the six column sums cannot be the same.

2. Each of 23 students is 10, 11, 12 or 13 years old, with at least one of each. Their total age is 253 years. How many 12 year olds are there if there are 1.5 times as many 12 year olds as 13 year olds?

   Solution:
   The total number of 12-year-olds and 13-year-olds is a multiple of 5, and is therefore one of 20, 15, 10 and 5. Suppose it is 10. Then their total age will be $6 \times 12 + 4 \times 13 = 124$. However, the total age of the other thirteen is at least 130, and cannot be $253 - 124 = 129$. If it is 15 or 20, the disparity is even greater. Hence there must be three 12-year-olds and two 13-year-olds, with total age $3 \times 12 + 2 \times 13 = 62$. The total age of the other eighteen is $253 - 62 = 191$. This is possible if the number of 11-year-olds is $191 - 18 \times 10 = 11$. In summary, seven are 10 years old, eleven are 11 years old, three are 12 years old and two are 13 years old.

3. On the line AB, 200 points are chosen symmetrically with respect to the midpoint of AB. Half of the points are red, and half are blue. Prove that the sum of the distances from A to all the red points is equal to the sum of the distances from B to all the blue points.

   Solution:
   Consider pairs of points symmetric with respect to the midpoint of AB. There are three kinds, red pairs, blue pairs and mixed pairs. Since half of the points are red, and half are blue, the number of red pairs is equal to the number of blue pairs. The contribution of each red pair to the A sum is AB. The contribution of each blue pair to the B sum is also AB. Hence these two kinds of pairs cancel each other. By symmetry, a mixed pair makes equal contributions to the A sum and to the B sum. The desired conclusion follows.

4. Seven genuine coins have the same weight. Two counterfeit coins also have the same weight. A counterfeit coin is heavier than a genuine coin. Identify the fake coins using a standard balance at most four times.

Solution:

Divide the nine coins into groups A, B and C, with three coins in each group. In the first two weighings, put A against B and C respectively. We cannot have equilibrium both times. Suppose A and B balance but C is lighter. Then there is a counterfeit coin in each of A and B. If C is heavier instead, then both counterfeit coins are in C. We cannot have B heavier than A and A heavier than C, nor can we have C heavier than A and A heavier than B. If we do not have equilibrium in either of the first two weighings, either A is heavy both times or light both times. In the former instance, both counterfeit coins are in A. In the latter instance, there is one counterfeit coin in each of B and C. With two weighing still available, we can identify the counterfeit coins, dealing with one group at a time.

5. Dissect a square into convex pentagons.

Solution:
One way is shown in Figure 2.

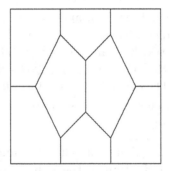

Figure 2

6. The map of a subway system is a convex polygon in which no three diagonals are concurrent. There is a station at each vertex and at every intersection of two diagonals. Train runs along entire diagonals, but not necessarily every diagonal. If each station lies on the route of at least one train, prove that it is possible to go from any station to any other station, changing trains at most twice.

Solution:
Let A and B be two arbitrary stations. Let A lie on a diagonal KL and B lie on a different diagonal MN. If these two diagonals intersect, we can get from A to B changing trains only at that intersection. Suppose the two diagonals do not intersect. We may assume that $KLMN$ is a convex quadrilateral, and there is a station at the intersection of KM and LN. Now some train must run on one of them, say KM. Then we can go from A to B changing trains only at K and M.

1. Adam and Betty wrote 54 tests marked out of 3. Checking their records, Adam had as many 3s as Betty had 2s, as many 2s as Betty had 1s, and as many 1s as Betty had 0s. Prove that their averages are not the same.

   Solution:
   In the tests where Adam had the higher score, he beat Betty by 1. In the tests where Betty had the higher score, she beat Adam by 3. If they had the same average, then Adam must have beaten Betty three times as often as Betty had beaten him. Hence the total number of tests must be a multiple of 4, and could not have been 54.

2. Does there exist a number consisting of each of the digits 1 to 9 exactly once such that between any two digits differing by 1, there are an odd number of other digits?

   Solution:
   Suppose that such a number exists. Paint the digital positions alternatively black and white. Then two consecutive digits must occupy positions of the same color, so that all nine digits occupy positions of the same color. This is clearly impossible.

3. Place nine points inside a triangle so that there are twelve points altogether. In addition to the three sides of the triangle, join some other pairs of these twelve points with non-intersecting segments so that each point is joined to five others, and the original triangle is divided into triangles.

   Solution:
   The configuration may be the net of an icosahedron, as shown in Figure 3.

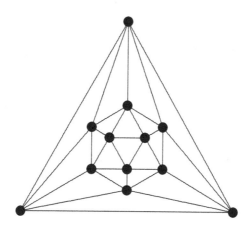

Figure 3

4. What is the smallest number of counters that must be placed on the squares of a 12×12 chessboard so that the shape in Figure 4 cannot be placed on three unoccupied squares of the chessboard?

Figure 4

Solution:
The shape in Figure 4 is called a V-tromino. Divide the chessboard into thirty-six 2×2 subboards. If we place less than 2 counters on a subboard, then there is enough room for the V-tromino. Hence we need at least 2 counters on each subboard, bring the total to $36 \times 2 = 72$. On the other hand, if we place a counter on each of the 72 black squares in the usual alternating color pattern, there is no room for the V-tromino.

5. Does there exist a positive integer such that its square begins with 123456789?

Solution:
Note that $111111111^2 = 12345678987654321$, as seen in the computation below, with trailing 0s omitted.

									1	1	1	1	1	1	1	1	1
×									1	1	1	1	1	1	1	1	1
1	1	1	1	1	1	1	1	1									
	1	1	1	1	1	1	1	1	1								
		1	1	1	1	1	1	1	1	1							
			1	1	1	1	1	1	1	1	1						
				1	1	1	1	1	1	1	1	1					
					1	1	1	1	1	1	1	1	1				
						1	1	1	1	1	1	1	1	1			
							1	1	1	1	1	1	1	1	1		
								1	1	1	1	1	1	1	1	1	
1	2	3	4	5	6	7	8	9	8	7	6	5	4	3	2	1

6. A certain country has only four kinds of bills, worth \$1, \$2, \$5 and \$10. Prove that from a stack of bills totaling \$400, an outsider can be paid exactly \$300.

Solution:

We claim that we can form 40 stacks of bills each totaling $10. Each $10 bill forms a stack. Every two $5 bills form a stack. Suppose there is one unmatched $5 bill. The total value of the remaining bills must be an odd multiple of 5. Thus they cannot all be $2 bills. We now add a $1 bill to the lone $5 bill. The total value of the remaining bills is now even. Thus we may treat two $1 bills as a single $2 bill. Our claim follows easily. To make up exactly $300, we just use 30 of the stacks.

1. A six-digit number is given. How many different seven-digit numbers have the property that if one digit is removed from it, the given six-digit number will be obtained?

   Solution:
   Before the first digit of the given six-digit number, we have exactly 9 choices for adding a digit because we cannot add 0. For each of the subsequent places for adding digits, we still only have 9 choices, because adding the same digit as the one in the preceding place produce a seven-digit number which has already been counted. So, we have 9 possibilities. Hence there are $9 \times 7 = 63$ such seven-digit numbers.

2. A grasshopper jumps 1 cm. Then it jumps 3 cm in the same or the opposite direction. Then it jumps 5 cm in the same or the opposite direction, and so on. Can the grasshopper get back to the starting point on the 25th jump?

   Solution:
   After making an odd number of jumps of an odd number of cm, the grasshopper must be at an odd number of cm from the starting point, and cannot be at the starting point.

3. The squares of a 5×5 chessboard are painted in one of two colors in an arbitrary way. Prove that there exist 2 rows and 2 columns such that the 4 squares where they intersect are all of the same color.

   Solution:
   Let the colors be red and green. A red column is one which contains at least 3 red squares, and a green column is one which contains at least 3 green squares. Since there are 5 squares in a column, each column is either red or green. By the Pigeonhole Principle, at least 3 of the 5 columns are of the same type, and by symmetry we may assume that they are red columns. Delete the other 2 columns. In the resulting 5×3 rectangle, there are at least 9 red squares. We consider two cases.
   Case 1. Some row consists of 3 red squares.
   If each of the other 4 rows contains at most 1 red square, then the total number of red squares will be less than 9. Hence another row contains at least 2 red squares. The desired conclusion follows.
   Case 2. Each row contains at most 2 red squares.
   Each of 4 rows contains 2 red squares. By Pigeonhole Principle again, 2 of these have red squares in the same columns. The desired conclusion follows.

4. Prove that if the sum of two positive integers is 770, then their product is not divisible by 770.

Solution:

Let n be one of the two positive integers. Then the other one is $770-n$. If 770 divides $n(770-n) = 770n - n^2$, then $770 = 2 \times 5 \times 7 \times 11$ divides n^2. It follows that $2 \times 5 \times 7 \times 11$ must divide n. However, n cannot be divisible by 770 since $0 < n < 770$. We have a contradiction.

5. Label the edges of a cube with 1, 2, 3, ..., 12 so that the sum of the labels of the four edges of each of the six faces is the same.

Solution:

Figure 5 shows one of many possible labelings.

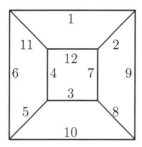

Figure 5

6. Anna and Boris play a game with a stack of 100 counters. Anna goes first, and moves alternate thereafter. In each move, a player divides a stack of at least two counters into two smaller piles. The loser is the player without a move, when each stack consists of exactly one counter. Prove that Anna must win this game.

Solution:

Each move increases the number of stacks by 1. At the beginning, there is only a single stack. At the end, there are 100 stacks. Hence 99 moves are made between the two players. Since 99 is odd and Anna makes the first move, she will also make the last move and win the game.

1. In a chess tournament, each participant plays every other participants exactly once. Each participant gets 1 point for a win, $\frac{1}{2}$ point for a draw and 0 points for a loss. At most how many of the 30 participants can score 18 points or more?

   Solution:
   A participant scoring 18 points or more is called a high achiever, and a lower achiever otherwise. We first suppose that there are 24 high achievers. Together, they must score at least $24 \times 18 = 432$ points. There are $\frac{24 \times 23}{2} = 276$ games played among them, and therefore 276 points to be scored. Even if they win every game against the 6 low achievers, they can only score an additional $24 \times 6 = 144$ points. However, $276 + 144 = 420$ points are not enough. We now show that we can have as many as 23 high achievers. If each of them draws with the other 22 and beats all 7 low achievers, the total score will be $11 + 7 = 18$ each.

2. Ten 10×20 chocolate pieces are cut up into twenty triangles. Pack them into a square box, leaving no empty space.

   Solution:
   The packing can be done as shown in Figure 6.

   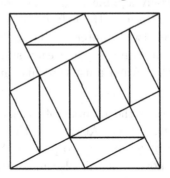

   Figure 6

3. Each of Benny, Denny, Kenny and Lenny either always lies or always tells the truth. Benny claims that Denny is a liar. Lenny claims that Benny is a liar. Kenny claims that both Benny and Denny are liars. Kenny also claims that Lenny is a liar. Which of them always lies and which of them always tells the truth?

Solution:

Since Benny claims that Denny is a liar, exactly one of them is a liar. Since Kenny claims that they are both liars, Kenny is himself a liar. Since he also claims that Lenny is a liar, Lenny always tells the truth. Since he claims that Benny is a liar, Denny is the other one who always tells the truth.

4. A solitaire game starts with eight numbers arranged in a circle. Each is either 1 or −1, and the choice is arbitrary. In each move, one can multiply any blocks of adjacent numbers of length 3 by −1. Prove that one can make all eight numbers equal to 1.

Solution:

Label the positions 1 to 8 in cyclic order. We always make moves five at a time. Suppose we change the signs of the numbers in (8,1,2), (2,3,4), (3,4,5), (5,6,7) and (6,7,8). The net change is that only the number in position 1 has its sign changed. Thus we can change the signs one at a time, and there is no difficulty in making all the −1s into 1s.

5. Baron Münchhausen has a time machine which allows a jump from March 1st to November 1st of any other year, from April 1st to December 1st, from May 1st to January 1st and so on. He cannot arrive and then depart on the same day. The Baron starts and ends his time travel on April 1st, and claims that he has been away for 26 months. Prove that he is mistaken.

Solution:

Let January, May and September be red, February, June and October be yellow, March, July and November be blue and April, August and December be green. If the Baron uses the time machine, he will arrive at a month of the same color but must wait a month before he can use it again. If he does not use it, he still waits a month. Hence the months he spends away must be red, yellow, blue and green in cyclic order. Since he departs and returns on a green month, the number of months he has been away must be a multiple of 4, and cannot be 26.

6. Four different digits are given. We use each of them exactly once to construct the largest possible four-digit number, and use each of them exactly once to construct the smallest possible four-digit number, which may not start with 0. If the sum of these two numbers is 10477, what are the given digits?

Solution:
We claim that if the digits of the two numbers are in reverse order, then the sum of the two numbers must be divisible by 11. The sum of the thousands digits of the two numbers is 1000 times the sum of their units digits, and $1000 + 1 = 1001$ is a multiple of 11. The sum of the hundreds digits of the two numbers is 10 times the sum of their tens digits, and $10 + 1 = 11$. This justifies our claim. Since 10477 is not divisible by 11, the digits of the two numbers are not in reverse order. This is only possible if one of the given digits is 0, which cannot be the thousands digit of the smaller number. It follows that 0 is the units digit of the larger number and the hundreds digit of the smaller number. Now the units digit of the smaller number is 7, which is therefore the thousands digit of the larger number. Hence the thousands digit of the smaller number must be 3 since no carrying over is coming this way. Then 3 is also the tens digit of the larger number. It follows that that the tens digit of the smaller number is 4, which is also the hundreds digit of the larger number. Thus the digits must be 7, 4, 3 and 0, and the addition is $7430 + 3047 = 10477$.

1. Prove that in the 400-digit number 84198419...8419 some digits can be deleted from the beginning and some from the end such that the sum of the remaining digits is equal to 1984.

   Solution:
   The sum of the digits of 8419 is 22. The sum of the digits of the 360-digit number 1984...1984 is 1980. We need to increase this sum by 4. Thus we delete the first digit 8 and the last thirty-eight digits 1984...198419, leaving behind the 361-digit number 41984...1984.

2. Construct a 4×4 table of non-zero numbers such that the sum of the numbers in the four corner squares of any 2×2, 3×3 or 4×4 subtable is 0.

   Solution:
   Figure 7 shows one of many possible constructions.

−1	2	−2	1
−2	1	−1	2
2	−1	1	−2
1	−2	2	−1

   Figure 7

3. On a line containing a segment AB, there are 45 points marked, none of which lie on the segment AB. Prove that the sum of the distances from these points to A is not equal the sum of their distances to B.

   Solution:
   For any of the 45 points, the difference between its distances from A and B is $\pm AB$. The sum of an odd number of copies of $\pm AB$ cannot be equal to 0.

4. Each square of an infinite chessboard is painted in one of 8 colors. Prove that it is possible to place a copy of the shape in Figure 8, possibly rotated or reflected, such that it covers two squares of the same color.

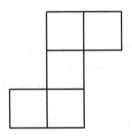

Figure 8

Solution:
By the Pigeonhole Principle, two squares in a 3×3 chessboard must have the same color. Any two squares in a 3×3 chessboard can be covered by a copy of the shape in Figure 8, possibly rotated or reflected.

5. A solitaire game starts with a circle divided into 6 sectors each containing a number as shown in Figure 9. In each move, one may add 1 to the numbers in any two adjacent sectors. Prove that it is impossible to make all six numbers equal.

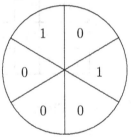

Figure 9

Solution:
Paint the sectors red and blue alternatively, with the two 1s in red sectors. The sum r of the numbers in red sectors is initially 2, while the sum b of the numbers in blue sectors is 0. In each move, both r and b increase by 1. Hence $r \neq b$ at any time. Thus it is impossible to make all six numbers equal.

6. Anna and Boris play a game with the numbers from 1 to 100 written in order in a row. Anna goes first, and turns alternate thereafter. In each move, a player puts one of the operation signs $+$, $-$ and $\times$ between any two numbers which do not already have an operation sign in between them. After 99 operation signs have been placed, the value of the expression is computed. Anna wins if this value is odd, and Boris wins if it is even. Prove that Anna has a winning strategy.

Solution:

Since only parity matters, we may replace all $-$s by $+$s. Let 0 represent an even number and 1 represent an odd number. Anna starts by adding the last two digits, in some alternating sequence which begins and ends in 1s. Suppose Boris inserts a $\times$ sign, necessarily between a 0 and a 1, generating two adjacent 0s. Anna inserts a $+$ sign between them. Suppose Boris inserts a $+$ sign, necessarily between a 0 and a 1, generating two adjacent 1s. Anna inserts a $\times$ sign between them. In either case, the sequence is shortened by two digits, but remains alternating. Moreover, it still begins and ends in 1s. Eventually, the sequence shortens to 101 and Anna wins.

1. Each of 68 coins has a different weight. In 100 weighings on a standard balance, find the heaviest and lightest coins.

   Solution:
   First weigh the coins in 34 pairs, and sort them into the heavy group and the light group. The heaviest coin must be in the heavy group. Using 33 weighings to eliminate one coin at a time, we can identify the heaviest coin. Similarly, 33 weighings in the light group will identify the lightest coin. The total number of weighings is 34+33+33=100.

2. A 45-digit number consists of one 1, two 2s, three 3s, ..., and nine 9s. Prove that it is not the square of an integer.

   Solution:
   The digit-sum of this number is $1^2 + 2^2 + \cdots + 9^2 = \frac{9 \times 10 \times 19}{6} = 3 \times 85$. It is divisible by 3 but not by 9, and cannot be the square of an integer.

3. A traveler departed from his home city A to the city B farthest from A. From B, he departed to the city C farthest from B, and so on. Prove that if C and A are different cities, then the traveler will never reach home.

   Solution:
   Suppose the traveler does not return to A directly from B. Then the length BC is greater than AB. The traveler cannot go from C to A as otherwise $CA > BC > AB$, and B would not have been her first destination. If she returns to B from C, she will travel forever between B and C. If she moves on to city D, then $CD > BC > AB$. Thus the length of the segments does not diminish, and her return to A would yield a contradiction as before.

4. Find 1000 numbers whose sum is equal to their product.

   Solution:
   Factoring the largest number under 1000, we have $999 = 27 \times 37$. Then $28 \times 38 - (28 + 38) = (28 - 1) \times (38 - 1) - 1 = 998$. If we take the numbers 28 and 38 and add 998 copies of 1, we will have 1000 numbers whose sum and product are both $28 \times 38 = 1064$.

5. Among 300 boots, 100 are of size 8, 100 of size 9 and 100 of size 10. There are 150 right boots and 150 left boots. Prove that one can select at least 50 pairs of boots each consisting of a right boot and a left boot of the same size.

Solution:

For each size, we have some matching pairs of boots and some unmatched boots. Suppose the total number of pairs of matching boots is less than 50. Then there are more than 200 unmatched boots. By symmetry, we may assume that all unmatched boots of size 8 or 9 are left boots while all unmatched boots of size 10 are right boots. Since the total number of left boots is equal to the total number of right boots, we have more than 100 unmatched right boots of size 10. This is a contradiction since there are only 100 boots of size 10.

6. A solitaire game starts with ten boxes numbered from 1 to 10 arranged randomly in two stacks. In each move, one may take several boxes from the top of a stack and put them on top of the other stack, which may be empty. Prove that one can form a stack with the ten boxes ordered from 1 at bottom to 10 at the top in 19 moves.

Solution:

We may assume that the stack containing box 1 is on space A, and the other stack is on space B. In the first move, put the entire stack on A on top of the stack on B. In the second move, put box 1 along with everything on top of it on A. In the third move, take off everything on top of box 1 and put them on top of the stack on B. In the fourth move, put box 2 along with everything on top of it on A. Thus every two moves puts a new box in the correct position in the stack on A. After 18 moves, the boxes 1 to 9 are in their correct positions. A 19th move, if necessary, will accomplish the task.

1. A solitaire game starts with the cards numbered 7, 8, 9, 4, 5, 6, 1, 2 and 3 placed in a row in that order. In each move, one may remove a block of any numbers of adjacent cards, reverse their order and put them back in the row. Show a sequence of moves which puts the cards in the order 1, 2, 3, 4, 5, 6, 7, 8 and 9.

   Solution:
   This can be accomplished in three moves as shown in Figure 10. The shaded cards do not move.

   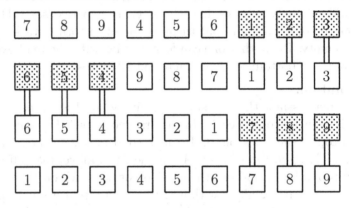

   Figure 10

2. There are 44 queens on a standard chessboard. Prove that each queen attacks at least one other queen.

   Solution:
   A queen attacks 7 squares horizontally, 7 vertically and at least 7 diagonally. If some queen does not attack any other queen, 21 squares must be empty. However, 44+21=65, and the chessboard has only 64 squares.

3. Let a and b be positive integers such that $34a = 43b$. Prove that $a + b$ is composite.

   Solution:
   Since, $34a = 43b$, $a = \frac{43}{34}b$ and so, $a + b = \frac{43}{34}b + b = \frac{77}{34}b$. Since 77 and 34 are relatively prime, $a+b$ must be divisible by 77. Since $77 = 7 \times 11$, $a + b$ is composite.

4. Find a configuration of several identical round coins on a table so that each coin touches exactly three other coins.

Solution:
Figure 11 shows four groups of four coins. Within each group, two of the coins touch all three others, while the other two touch only two others. These coins at the end all touch similar coins from other groups.

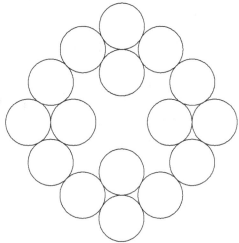

Figure 11

5. There are 55 numbers placed on a circle. Each number is the sum of its neighbors. Prove that all numbers are equal to zero.

   Solution:
   Let the numbers be $a_1, a_2, \ldots, a_{55}$. Then we have $a_3 = a_2 - a_1$ and $a_4 = a_3 - a_2 = -a_1$. Hence $a_7 = a_1$, and it follows that $a_{55} = a_1$. Since $a_1 = a_{55} + a_2$, we must have $a_2 = 0$. Since we can choose any of the 55 numbers to be a_1, all numbers are equal to 0.

6. (a) Find a seven-digit number with distinct digits which is divisible by all of its digits.

   (b) Does their exist an eight-digit number with this property?

   Solution:
   We answer the two parts in reverse order.

   (b) Clearly, our number cannot contain the digit 0. Since it must contain an even digit, it cannot contain the digit 5. Thus it must contain all of the digits 1, 2, 3, 4, 6, 7, 8 and 9. Their sum is 40, and no such number can be divisible by 9.

   (a) The largest multiple of 9 below 40 is 36. So we omit the digit 4. Our number is automatically divisible by 1, 3 and 9. If the last digit is even, it will be divisible by 2 and 6. A little testing reveals that 1369872 is also divisible by 7 and 8.

1. A solitaire game starts with the 4×4 table in Figure 12 on the left. In each move, one may add 1 to all the numbers in any row or subtract 1 from all the numbers in any column. How can one obtain the table in Figure 12 on the right?

1	2	3	4
5	6	7	8
9	10	11	12
13	14	15	16

1	5	9	13
2	6	10	14
3	7	11	15
4	8	12	16

Figure 12

Solution:
We first add 1 to the first row 9 times, to the second row 6 times and to the third row 3 times, resulting in the table in Figure 13. It is symmetric about the main diagonal. To reach the target table, we just subtract 1 from the first column 9 times, from the second column 6 times and from the third column 3 times.

10	11	12	13
11	12	13	14
12	13	14	15
13	14	15	16

Figure 13

2. A certain country has only four kinds of bills, worth $1, $10, $100 and $1000. Can exactly half a million bills be worth exactly one million dollars?

Solution:
Suppose we devalue each bill by $1. Since we have half a million bills, the total value has been reduced by half a million dollars. Now every bill is worth a number of dollars equal to a multiple of 9, but half a million is not a multiple of 9. Hence the situation is impossible.

3. Each of six cities is connected to the others by five roads. Show that it is possible for the roads to intersect only three times with exactly two roads crossing over at each intersection. Junctions at the cities are not considered intersections.

Solution:
A possible arrangement is shown in Figure 14.

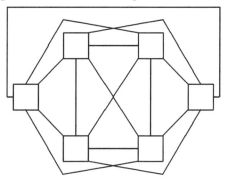

Figure 14

4. The cost of a hotdog and the cost of a burger are both integral numbers of cents. If each boy buys a hotdog and each girl buys a burger, their total expenditure will be one cent more than if each boy buys a burger and each girl buys a hotdog. There are more boys than girls. How many more?

Solution:
The change in total expenditure is a multiple of the difference between the numbers of boys and girls. Since this change is 1 cent, the number of boys exceeds the number of girls by 1.

5. A six-digit number from 000000 to 999999 is said to be *lucky* if the sum of its first three digits is equal to the sum of its last three digits. What is the minimum length of a block of consecutive numbers which will guarantee the inclusion of at least one lucky number, whatever the first number of the block may be?

Solution:
The first lucky number is 000000. The second one is 001001. If the length of the block is at most 1000 and the first number of the block happens to be 000001, there will not be a lucky number in the block. We now prove that a length of 1001 is sufficient. A six-digit number is said to be *fortunate* if its first digits are the same as its last three digits. Hence all fortunate numbers are lucky. The difference between one fortunate number and the next is 1001. The desired conclusion follows immediately.

6. Anna and Boris play a game on a 9×9 chessboard. Anna goes first and turns alternate thereafter. In each move, Anna puts a red counter on a vacant square while Boris puts a blue counter on a vacant square. When the board is completely filled, a row with more red counters than blue counters is called a red row, and a blue row otherwise. Red and blue columns are similarly defined. The score for Anna is the sum of the numbers of red rows and red columns while that for Boris is the sum of the numbers of blue rows and blue columns. What is the highest possible score for Anna?

Solution:

Divide the chessboard into the central square and forty dominoes as shown in Figure 15. We first show that Anna can score as high as 10 points. She starts by putting a red counter in the central square. Thereafter, she puts a red counter in the same domino where Boris has just put a blue counter. She will score row 5 and column 5, one of rows 1 and 2 and one of columns 1 and 2, one of rows 3 and 4 and one of columns 3 and 4, one of rows 6 and 7 and one of columns 6 and 7, as well as one of rows 8 and 9 and one of columns 8 and 9.

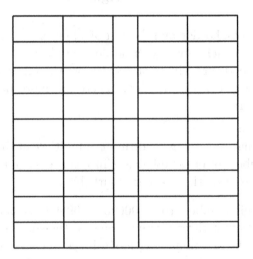

Figure 15

We now show that Boris can hold Anna down to 10 points. Whenever she puts a red counter in a domino, he puts a blue counter in the same domino. If Anna puts a red counter in the central square, Boris makes an arbitrary move. If Anna completes a domino he has started, Boris makes an arbitrary move except in the central square. Eventually, Anna must put a red counter in the central square. The scoring is exactly the same as before.

1. A solitaire game starts with a 0 in each square of a 3×3 table. In each move, one may add 1 to all numbers in any of the four 2×2 subtables. Is it possible to obtain the table in Figure 16?

4	9	5
10	18	12
6	13	7

Figure 16

Solution:
Each 2×2 subtable contains the central square and exactly one of the corner squares. Hence after any number of moves, the number in the central square is always equal to the sum of the numbers in the four corner squares. Since $18 \neq 4 + 5 + 6 + 7$, the task is impossible.

2. A teacher has a deck of 30 cards numbered from 1 to 30, as does each of 30 students. They all turn over the top cards in their respective decks. Whenever the number on a student's card matches the number on the teacher's card, that student scores 1 point. When all the cards have been turned over, each student scores a different number of points. Prove that one of them scores 30 points.

Solution:
It is impossible for a student to score exactly 29 points. Hence the 30 different scores must be 0, 1, 2, ..., 27, 28 and 30.

3. Is it possible to arrange the positive integers from 1 to 100 inclusive in a row so that the difference between any two adjacent numbers is at least 50?

Solution:
A possible arrangement is 51, 1, 52, 2, 53, 3, ..., 99, 49, 100, 50.

4. Do there exist non-zero integers such that one of them is divisible by their sum and the other is divisible by their difference?

Solution:
If two integers are both positive or both negative, the absolute value of their sum exceeds the absolute value of either of them, so that neither is divisible by the sum. If one is positive and the other is negative, the absolute value of their difference exceeds the absolute value of either of them, so that neither is divisible by the difference. It follows that such two integers do not exist.

5. A solitaire game starts with a pile of 1001 counters. In each move, a pile with at least three counters is chosen. One counter is removed while the remaining counters are split into two piles, not necessarily equal in size. Is it possible that after a number of moves, each remaining pile contains exactly three counters?

Solution:
Consider the sum of the number of piles and the number of counters. This sum is initially $1001 + 1 = 1002$. In each move, a new pile is created while a counter is removed. So the value of the sum remains unchanged. If the task is possible, then the sum must be divisible by 4. However, 1002 is not.

6. A solitaire game starts with an 8×8 chessboard in which all 64 squares are white. In each move, one may enter the chessboard on a border square, visit parts of the chessboard going between two squares with a common side, and exit via a border square. The color of a square is changed from white or black or vice versa every time it is visited. Is it possible to create the usual alternating color pattern on the chessboard?

Solution:
It is possible to create any pattern since one can change the color of exactly one square in any move. Taking any path to visit that square, one then exists along the same path. Every other square is visited an even number of times, and its color is the same as before.

1. Each of the competition for the fifth, sixth, seventh, eighth, ninth and tenth grades consists of seven problems. In each competition, exactly four of the questions do not appear on the competition of any other grade. What is the greatest number of distinct problems that may appear in these six competitions?

   Solution:
   The four problems in each competition which are not duplicated yield a total of $6 \times 4 = 24$ distinct problems. To maximize the number of distinct problems, each of the duplicated problems must appear exactly twice, yielding an additional $6 \times 3 \div 2 = 9$ distinct problems, for a total of $24+9 = 33$. This is easily arranged if the competitions for the fifth and sixth grades have three common problems, the competitions for the seventh and eighth grades have three common problems, and the competitions for the ninth and tenth grades have three common problems.

2. Prove that among the six-digit numbers from 000000 to 999999, there are as many with digit sum 27 as those where the sum of its first three digits is equal to the sum of its last three digits.

   Solution:
   Two digits are said to be complements of each other if their sum is 9. Consider a six-digit number with digit sum 27. If we replace each of the first three digits with its complement, we obtain a six-digit number where the sum of its first three digits is equal to the sum of its last three digits. This replacement process is reversible, and the desired conclusion follows from the one-to-one correspondence.

3. In a model railway set, there are two kinds of track pieces as shown in Figure 17. They may not be flipped over. When two pieces are put together, the convex head of one must fit into the concave tail of the other. A closed track constructed according to this rule has been taken apart, and one piece has been replaced by a piece of the other kind. Prove that it is now impossible to reassemble the pieces to form a closed track which follows the rule.

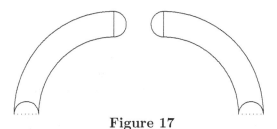

Figure 17

Solution:

Put an arrow along the neck of each piece as shown in Figure 18. Suppose we move along a legal closed track, and record the changing direction of the arrow as we go from piece to piece. Going through a clockwise piece, the arrow makes a 90° turn clockwise. Going through a counterclockwise piece, the arrow makes a 90° turn counterclockwise. When we return to the starting piece, the arrow is in its original direction. This means that some of the clockwise turns cancel out some of the counterclockwise turns. The difference between the numbers of these two types of turns must be a multiple of $360° \div 90° = 4$. Thus the difference between the numbers of the two kinds of pieces is also a multiple of 4. When a piece is replaced by a piece of the other kind, this difference will be an even number not divisible by 4. Hence a closed track cannot be reassembled legally.

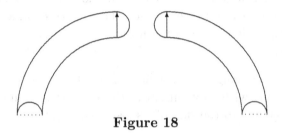

Figure 18

4. Each of 32 coins has a different weight. In 35 weighings on a standard balance, find the heaviest and the second heaviest coins.

Solution:

First weigh the coins in 16 pairs and record the result of each weighing. Set aside the 16 coins in the light group. Then weigh the coins in the heavy group in 8 pairs and record the result of each weighing. Continue in this until all but the heaviest coin has been set aside. We have used $16 + 8 + 4 + 2 + 1 = 31$ weighings so far. Now the second heaviest coin must be one of the five eliminated by the heaviest coin, and it takes 4 more weighings to eliminate four of them.

5. Find two six-digit numbers such that the twelve-digit number obtained when the digits of one are written after the digits of the other is divisible by the product of the original six-digit numbers.

Solution:

Since the six-digit number in front divides itself as well as the twelve-digit number, it must also divide the six-digit number at the back. The quotient of this division is a single digit d such that when it is added to 1000000, the sum is divisible by the six-digit number at the back. Now $1000001 = 101 \times 9901$ is not helpful, but $1000002 = 2 \times 3 \times 166667$ leads to 166667333334. Indeed, $\frac{166667333334}{166667 \times 333334} = \frac{1000002}{333334} = 3$.

6. Anna and Boris play a game on a 10×10 chessboard. Anna goes first, and turns alternate thereafter. In each move, a player puts either a red counter or a green counter on a vacant square. A player wins by completing a block of three adjacent counters of the same color along a row, a column or a diagonal. Which player, if either, has a winning strategy?

Solution:
Boris has a winning strategy. Whenever Anna offers him a position from which he can win the game immediately, he will of course make the winning move. If not, he just places a counter different in color to the one Anna has just played, and into the square symmetric about the center of the board to the square where Anna has just placed her counter. Boris will never offer Anna a winning position, since by symmetry Anna would have to offer him one in her last move. It follows that Boris cannot lose. We now prove that he must win. Suppose to the contrary that the board is completely filled without producing a winner. Figure 19 shows the central 4×4 subboard when there are two counters in the central 2×2 subboard, necessarily of different colors, and Anna has just put a red counter into square f6.

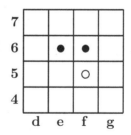

Figure 19

Note that square g6 cannot contain a red counter as otherwise Boris would have put a red counter in square f6 and wins. If it is vacant, Boris wins by putting a red counter there. Suppose it contains a green counter. Obviously, square e4 cannot contain a green counter. It cannot be vacant either as otherwise Boris would have put a green counter there and wins. If it contains a red counter, Boris would also have won by putting a red counter in e5. We have a contradiction.

95

1. Paula numbers the 96 sheets in her notebook page by page in order from 1 to 192. Nick rips out 25 sheets at random and adds together all 50 page numbers. Prove that his sum cannot be equal to 1990.

   Solution:
   The sum of the two page numbers on the same sheet must be odd. The sum of 25 odd numbers is still odd and cannot be equal to 1990.

2. Among 101 coins, 100 are genuine and have the same weight. The weight of the counterfeit coin is different from that of a genuine coin. Determine whether the counterfeit coin is heavier or lighter in two weighings on a standard balance. The identification of the counterfeit coin is not required.

   Solution:
   First, weigh 50 coins against 50 other coins. If we have equilibrium, the coin left out is counterfeit, and a second weighing will settle the issue. Suppose we do not have equilibrium in the first weight. Then the counterfeit coin is in one of the pans. We weigh 25 coins on the heavier pan against the other 25 coins on the same pan. If we have equilibrium, the counterfeit coin is lighter. Otherwise, it is heavier.

3. Is it possible to pack thirty-nine 5×11 chocolate pieces into a 39×55 box?

   Solution:
   Suppose the task is possible. Then the side of the box of length 39 is divided into segments of length 5 or 11. The number of segments of length 11 is 0, 1, 2 and 3, leaving respectively lengths of 39, 28, 17 and 6 to be made up with segments of length 5. However, none of these four numbers is divisible by 5. We have a contradiction.

4. Anna and Boris play a game starting with the number 1234. Anna goes first, and turns alternate thereafter. In each turn, the player subtracts from the number one of its non-zero digits. A player wins if the number is reduced to 0. Who has a winning strategy, Anna or Boris?

   Solution:
   Anna winning opening move is $1234 - 4 = 1230$. Her general strategy is to leave for Boris a number ending in 0. If the number is already 0, she has won. Otherwise, there is at least one non-zero digit before the final 0. Boris is forced to subtract a non-zero digit, leaving behind a number not ending in 0. So Anna can subtract the last digit and continue with her winning strategy.

5. Together, three students solved exactly 100 problems. Separately, each of them solved exactly 60 problems. A problem was considered difficult if it was solved by only one of them, and easy if it was solved by all of them. Prove that the number of difficult problems exceeded the number of easy problems by 20.

Solution:
Note that $3 \times 60 - 100 = 80$. If the 80 duplicate solutions are all to different problems, there are 0 easy problems and $100 - 80 = 20$ hard problems. Indeed $20 - 0 = 20$. Every easy problem takes away one duplicate solution and creates a hard problem. Since the number of easy problems and the number of hard problems change by the same amount, the difference remains 20.

6. For any boy in a club, all the girls who know him know one another. Each girl knows more boys than girls other than themselves. Prove that there are at least as many boys as girls in this club.

Solution:
Choose some girls to form an executive committee as follows. Start with an arbitrary girl. A girl is added to the committee if she does not know any girl already on the committee. Eventually, no further addition is possible. Then no two girls on the committee know each other, but every other girl knows at least one of them. Each girl on the committee writes down the names of all the boys and girls she knows, along with her own name. Since she knows more boys than girls other than herself, her list contains as many boys' names as girl's names. A boy's name cannot be on two lists as the girls who list him do not know each other. It follows that the total number of boys is at least the sum of the numbers of names of boys on all the lists. On the other hand, the total number of girls is at most the sum of the numbers of girls on all the lists. Hence there are at least as many boys as girls in the club.

1. Each of 40 children in a workshop class has nails, nuts and bolts. There are exactly 15 children with unequal numbers of nails and nuts, and 10 children with equal numbers of nuts and bolts. Prove that there are at least 15 children with unequal numbers of nails and bolts.

   Solution:
   Apart from the 15 children with unequal numbers of nails and nuts, the other $40 - 15 = 25$ have equal numbers of nails and nuts. At most ten of them can have equal numbers of nuts and bolts. Hence at least $25 - 10 = 15$ children have unequal numbers of nails and bolts.

2. A strange rule in a club allows the children only to trade any two marbles for any three others, or any three marbles for any two others. Is it possible to start with 100 red marbles and end up with 100 green marbles, having traded away exactly 1991 marbles in the meantime?

   Solution:
   Suppose we end up with 100 marbles. Since the total number of marbles has not changed, we must have traded two for three just as often as three for two. In each such pairs of trade, we have traded away 5 marbles. Since 1991 is not a multiple of 5, the situation is impossible.

3. Four girls are starting simultaneously from the same point on a circular track, running at constant speeds which are not necessarily the same. A and B run clockwise while C and D run counterclockwise. A meets C for the first time at the same moment as B meets D for the first time. Prove that A catches up with B for the first time at the same moment as D catches up with C for the first time.

   Solution:
   At the first time when A meets C and D meets B, each pair has covered the circular track exactly once. Hence the total speed of A and C is equal to the total speed of D and B. It follows that the difference between the speeds of A and B is the same as the difference between the speeds of D and C. This is equivalent to the desired conclusion.

4. Baron Münchhausen hunts ducks every day. One day, he declares, "Today I will bring home more ducks than two days ago but fewer than one week ago." For at most how many consecutive days can the Baron say this without telling a lie?

   Solution:
   We first show that the Baron can make true declarations from Day 8 to Day 13 in the chart below.

Days	1	2	3	4	5	6	7	8	9	10	11	12	13
Ducks	9	9	9	9	9	5	1	6	2	7	3	8	4

We now prove that at least one declaration by the Baron in a seven-day period is a lie. Suppose he can make true declarations from Day 3 to 9. We use the notation $i \to j$ to denote the statement that the number of ducks brought home on Day i is more than the number of ducks brought home on Day j. Then we have

$$1 \to 8 \to 6 \to 4 \to 2 \to 9 \to 7 \to 5 \to 3 \to 1.$$

This is a contradiction since a chain of inequalities cannot close up into a cycle.

5. Anna and Boris play a game with a red stick, a white stick and a blue stick, each of which is 1 meter long. Anna starts by breaking the red stick into three pieces. Then Boris breaks the white stick into three pieces. Finally, Anna breaks the blue stick into three pieces. She wins if she can use the nine pieces to form three triangles with sides of different colors. Can Boris stop her from winning?

Solution:
Boris cannot stop Anna from winning if she breaks each of her sticks into pieces of lengths $\frac{1}{2}$ meter, $\frac{1}{4}$ meter and $\frac{1}{4}$ meters. No matter how Boris breaks his stick, the longest piece will be shorter than 1 meter and each of the other two pieces will be shorter than $\frac{1}{2}$ meter. The longest piece of Boris will be combined with the two pieces of length $\frac{1}{2}$ meter while each of the other two piece of Boris will be combined with two pieces of length $\frac{1}{4}$ meters of different colors. In each case, we have an isosceles triangle in which the length of the base is less than the total length of the two arms.

6. In a tournament without draws, every two of the nine teams play against each other exactly once. Must there always be two teams such that every other team has lost to either or both of them?

Solution:
Such two teams need not exist. Let the teams form three divisions, each with three teams. Within each division, team 1 beats team 2, team 2 beats team 3 and team 3 beats team 1. Across divisions, every team in division 1 beats every team in division 2, every team in division 2 beats every team in division 3, and every team in division 3 beats every team in division 1. Suppose two teams with the desired property exist. They cannot be in the same division as both will lose to every team in another division. By symmetry, we may assume that they are in divisions 1 and 2. Then the team in division 1 loses to another team within the division, and that team also beats the team in division 2.

1. In a tournament, each participant plays every other participants exactly once. Each participant gets 1 point for a win, 0 points for a draw, and −1 point for a loss. One of the participants finishes the tournament with 7 points and another with 20. Prove that there is at least one drawn game.

   Solution:
   The participant with 20 points has 20 wins more than losses, while the participant with 7 points has 7 wins more than losses. Suppose there are no tied games. Then the former has played an even number of games while the latter has played an odd number of games. This is a contradiction since every participant plays the same number of games.

2. In a heptagonal castle, each of the seven sides is a straight wall and there is a watchtower at each of the seven vertices. The guards stay in the watchtowers. Each guard watches over both walls meeting at that watchtower. What is the minimum number of guards required so that each wall is watched over by at least 7 guards?

   Solution:
   A guard-wall pair consists of a guard and the wall he is watching over. Each of the seven walls appears in at least 7 guard-wall pairs, while each guard appears in exactly 2 guard-wall pairs. Hence at least $\lceil \frac{7}{2} \rceil = 25$ guards are required. It is sufficient to have 25 guards. In cyclic order, the numbers of guards in the 7 watchtowers are 4, 3, 4, 3, 4, 3 and 4.

3. Adam and Betty are of the same age. Adam multiplies his age this year by his age last year. Betty calculates the square of her age next year. Prove that the two answers have different digit-sums.

   Solution:
   The digit-sum of a multiple of 3 is divisible by 3, while the digit-sum of a non-multiple of 3 is not divisible by 3. We claim that one of the two answers is a multiple of 3 while the other is not. The desired conclusion then follows. Our claim is justified by considering the remainders when numbers are divided by 3.

Case Number	Last Year	This Year	Next Year	Adam's Answer	Betty's Answer
I	2	0	1	0	1
II	0	1	2	0	1
III	1	2	0	2	0

4. Fyodor collects coins. No coin in his collection is more than 10 cm in diameter. He keeps all the coins arranged side by side in a rectangular box of size 30 cm by 70 cm. Prove that he can fit all of his coins in another rectangular box of size 40 cm by 60 cm.

Solution:
Divide the 30×70 box into two 30×40 parts, with an overlap of width 10, as shown in Figure 20 on the left. Divide the 60×40 box into two 30×40 halves, as shown in Figure 20 on the right. Since no coin has diameter exceeding 10, each coin lies entirely within one of the two parts of the original box, and can be assigned to it. If we transfer all the coins from each part of the original box in exactly the same formation into a different half of the new box, everything will fit.

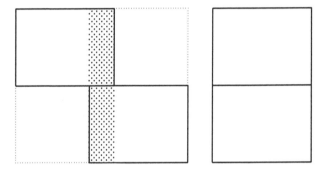

Figure 20

5. A circle is divided into 27 equal arcs by 27 points. Each point is either white or black. No two black points are adjacent or separated by only one white point. Prove that three of the white points are the vertices of an equilateral triangle.

Solution:
There are 9 sets of points that are the vertices of an equilateral triangle. If at least one point in each set is black, then there are at least 9 black points and at most 18 white points. Since there are at least two white points between two black points, there are exactly 9 black points, with exactly two white points between two of them. Now the black points form 3 of the 9 sets. Thus any of the other 6 sets consists of three white points which are the vertices of an equilateral triangle.

6. Three counterfeiters print bills of arbitrary integral denominations. Each one prints bills totaling $100, and can pay either of the other two counterfeiters any amount up to $25, perhaps with change. Prove that together they can pay an outsider exactly any amount from $100 to $200.

Solution:

We first prove an auxiliary result.

Lemma.

There is an amount between $25 and $50 such that at least one counterfeiter has a number of bills with total value of exactly that amount.

Proof:

Let one counterfeiter pay $25 to a second counterfeiter, perhaps with change. If an amount between $25 and $50 is paid by the first counterfeiter, there is nothing further to prove. If it is between $50 and $75, then the bills of the first counterfeiter not involved in this transaction have total value between $25 and $50. Finally, if the amount paid by the first counterfeiter is between $75 and $100, then the second counterfeiter gives change between $50 and $75. The bills of the second counterfeiter not involved in this transaction have total value between $25 and $50.

We tackle the main problem by considering three cases, according to the range of the amount to be paid to the outsider.

Case 1. The amount is between $100 and $125.

Subtract $100 from the desired amount. The difference is between $0 and $25. Let one counterfeiter pay the difference to a second counterfeiter, perhaps with change. Now the second counterfeiter can pay the outsider exactly the desired amount.

Case 2. The amount is between $125 and $150.

We may assume that the third counterfeiter has the exact amount specified by the Lemma. Subtract $100 from the desired amount. If the first difference is greater than or equal to the amount in the Lemma, subtract that amount from the first difference. Let the first counterfeiter pay the second difference, which is between $0 and $25, to the second counterfeiter, perhaps with change. Now the second counterfeiter can pay the outsider exactly $100 plus the first difference, and the third counterfeiter can pay the balance. If the first difference is less than the amount in the Lemma, subtract the first difference from this amount. Let the second counterfeiter pay the second difference, which is between $0 and $25, to the first counterfeiter, perhaps with change. Now the second counterfeiter can pay the outsider exactly $100 minus the first difference, and the third counterfeiter can pay the balance.

Case 3. The amount is between $150 and $200.

Subtract this amount from $300, so that the difference is between $100 and $150 inclusive. By Case 1 or 2, this difference can be paid to an outsider exactly. Setting aside the bills involved in this transaction, the remaining bills will have total value equal to the desired amount.

Chapter Four: Look Back

Set A : Problems on Differences

All four problems are based on the operation of subtraction, the computation of differences. A practical scenario is the measurement of length. Suppose we align a marked ruler against a line segment. If the left endpoint corresponds to the marking of 3 cm and the right endpoint corresponds to the marking of 7 cm, then the length of the segment is $7 - 3 = 4$ cm. We consider a related problem.

The proposer, **Solomon Golomb**, is arguably the greatest American-born scientist. His seminal research on shift register sequences is pivotal to the transition from analog computers to digital computers. His fingerprints are all over mathematics, computing science and electrical engineering.

There is also a lighter side to this intellectual giant. Golomb is the leader of the polyomino cult (see Set C), with his book *Polyominoes* as the bible. He was a regular contributor to the famous column *Mathematical Games* of **Martin Gardner** in *Scientific American*.

In one of these columns, Golomb pointed out that the ordinary ruler is inefficient, in that it can do without some of its markings. For instance, with a ruler of length 3 cm, the marking of 2 cm is redundant. All we need are the markings of 0 cm, 1 cm and 3 cm, because we can get 2 from $3 - 1$. With a ruler of length 6 cm, we can eliminate the markings of 2 cm, 3 cm and 5 cm, because $1 = 1 - 0$, $2 = 6 - 4$, $3 = 4 - 1$, $4 = 4 - 0$, $5 = 6 - 1$ and $6 = 6 - 0$.

Consider now a ruler of length 10 cm. Can we keep just three markings in addition to the markings of 0 cm and 10 cm? With five markings, we can measure up to ten distances. Ideally, these would be from 1 cm to 10 cm. However, no matter how the three extra markings are chosen, at least two of the ten distances are identical, and an ideal ruler of length 10 cm does not exist.

The closing of one problem usually opens up another problem. Suppose we have a ruler with three extra markings in addition to the two endpoints. What is the minimum length of the ruler so that the ten measurable distances are all distinct? It turns out that a length of 11 cm is sufficient. The readers are invited to explore rulers of minimum lengths with four of more extra markings, so that all measurable distances are distinct.

© The Editor(s) (if applicable) and The Author(s), under exclusive license to Springer Nature Switzerland AG 2020
K. Garaschuk, A. Liu, *Grade Five Competition from the Leningrad Mathematical Olympiad*,
Problem Books in Mathematics, https://doi.org/10.1007/978-3-030-52946-8_4

Set B : Parity

When an integer is divided by 2, the remainder is either 0 or 1. In the former case, we say that the integer is an *even* number. In the latter case, we say that the integer is an *odd* number. The *parity* of an integer is its status of being even or odd. Clearly, an integer cannot be both even and odd. It is surprising how powerful such an innocuous statement can be. It is the underlying idea to the solution of all three problems in this set.

More generally, any classification of objects into two types may be considered a parity situation. In everyday life, the two classes are usually in sharp contrast with each other, such as good and evil, cheap and expensive, and so on. This is sometimes summarized as the *ying* and the *yang* of the universe.

Here is a problem which does not at first seem to have any relationship to parity.

There are three soccer balls on a field. A girl kicks one of them so that it goes across the line joining the other two. She does this 25 times, kicking the balls in no particular order. Is it possible for all three balls to end up on their respective starting spots?

One immediately suspects that the task is impossible because 25 is an odd number. However, we need to explain further why this is so. Label the balls A, B and C, so they form a triangle. Reading the vertices of the triangle clockwise and starting with A, we may either get ABC and ACB. These are the two orientations of the triangle. Every kick, no matter which ball is kicked, as long as it crosses the line joining the other two, changes the orientation of the triangle. Suppose the initial orientation is ABC. After 25 kicks, the orientation has changed to ACB. So it is not possible for all three balls to return to their respective starting spots.

The arithmetic of parity is very straight-forward. The sum of two odd integers is even, as is the sum of two even integers. On the other hand, the sum of an odd integer and an even integer is odd. The product of two odd integers is odd, and the product of an even integer with any integer is even.

Parity is a prototype of *modular arithmetic*. Since we are dividing by 2, the *modulus* of this arithmetic is 2, and it is called *arithmetic modulo 2*. There are only two numbers in this arithmetic, namely 0 and 1. Actually, each of 0 and 1 represents a whole class of numbers.

- The number 0 represents the even integers $\{\ldots, -4, -2, 0, 2, 4, \ldots\}$.

- The number 1 represents the odd integers $\{\ldots, -3, -2, 1, 3, \ldots\}$.

Because the representatives 0 and 1 are the remainders in the division by 2, modular arithmetic is also known as *remainder arithmetic*.

The earlier statement that the sum of two odd integers is even can now be expressed simply as $1 + 1 = 0$. Similarly, we have $0 + 0 = 0$, $1 + 0 = 1$, $1 \times 1 = 1$, $0 \times 0 = 0$ and $0 \times 1 = 0$.

The numbers in arithmetic modulo 3 are 0, 1 and 2.

- The number 0 represents the multiples of 3.

- The number 1 represents numbers 1 more than multiples of 3.

- The number 2 represents numbers 1 less than multiples of 3.

The arithmetic can be summarized in the following operation tables.

+	0	1	2
0	0	1	2
1	1	2	0
2	2	0	1

×	0	1	2
0	0	0	0
1	0	1	2
2	0	2	1

In arithmetic modulo 4, the numbers are 0, 1, 2 and 3. The operation tables are

+	0	1	2	3
0	0	1	2	3
1	1	2	3	0
2	2	3	0	1
3	3	0	1	2

×	0	1	2	3
0	0	0	0	0
1	0	1	2	3
2	0	2	0	2
3	0	3	2	1

Set C : The Pigeonhole Principle

The Pigeonhole Principle, along with the concept of parity, are the two most important ideas in elementary mathematics. This result can be stated in the following colorful form.

The Pigeonhole Principle.
If there are more pigeons than holes and all the pigeons go into the holes, then there is at least one hole with at least two pigeons.

The proof is straightforward. If the result is not true, then every hole has at most one pigeon. Then the number of pigeons is at most equal to the number of holes, contrary to the assumption that it is greater. The versatility of this result lies in the fact that the pigeons and holes can represent almost anything.

When fractions are expressed as equivalent decimals, the expansions either terminate as in $\frac{1}{2} = 0.5$ and $\frac{2}{25} = 0.08$ or repeating, as in $\frac{1}{3} = 0.33\ldots$ and $\frac{1}{7} = 0.142857142857\ldots$. Take for example the conversion of $\frac{1}{7}$. We divide the numerator by the denominator and extend the division beyond the decimal point. In each step, the remainder is one of 0, 1, 2, 3, 4, 5 or 6. If 0 appears as a remainder, the decimal expansion terminates. If not, then by the seventh step, we would have seven remainders (seven pigeons) each of which is one of 1, 2, 3, 4, 5 and 6 (six holes). Then two of the remainders will have to be the same, and the decimal expansion will repeat.

In Problem 1, the unmatched boots of each size are either all left boots or all right boots. Let three pigeons represent boots of sizes 8, 9 and 10 respectively. A pigeon goes into the first hole if the unmatched boots of that size are left boots, and into the second hole if they are right boots. By the Pigeonhole Principle, there is a hole with two pigeons. By symmetry, we may assume that the unmatched boots of sizes 8 and 9 are left boots, as claimed in the solution to Problem 1.

In Problem 2, after placing a counter on each square in the top row, we are left with 19 counters and a 6×6 chessboard. The 19 counters are the pigeons and the nine 2×2 subboard into which the chessboard is divided are the holes. We need a more general result.

The Generalized Pigeonhole Principle.
If all the pigeons go into the holes, then there is at least one hole with at least the average number of pigeons per hole.

This will be given a more appropriate name in the next Section. Note that the original result follows immediately. If there are more pigeons than holes, the average number of pigeons per hole is greater than 1. Since the pigeons remain intact, this means that there is at least one hole with at least two pigeons.

Returning to Problem 2, the average number per hole is $19 \div 9 > 2$. Hence some subboard must contain at least three counters. Note that 25 is the minimum. With only 24 counters, we can use them to fill the first, the third, the fifth and the seventh row, and no 2×2 subboard will contain three counters.

In Problem 3, we seek to bar a certain shape from a chessboard. Let the chessboard be infinite. Shapes formed of unit squares joined edge to edge are named *polyominoes* by **Solomon Golomb**. In addition to the monomino and the domino, there are 2 trominoes, 5 tetrominoes and 12 pentominoes. They are shown in Figure 1. The letter names for the trominoes, tetrominoes and the pentominoes are suggested by their shapes.

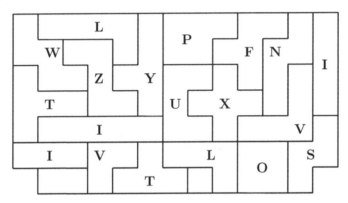

Figure 1

We wish to determine the fraction of squares that must be blocked off in order to bar a specific polyomino from the infinite chessboard. For the monomino, this fraction is trivially 1. For the domino and the V-tromino, it is $\frac{1}{2}$. For the I-tromino, it is $\frac{1}{3}$. Since the area of a tetromino is 4, the minimum fraction for a tetromino is at least $\frac{1}{4}$. The arrangement in Figure 2 on the left shows that it is exactly $\frac{1}{4}$ for the I-tetromino, while the arrangement in Figure 2 on the right shows that it is so for the O-tetromino and the S-tetromino.

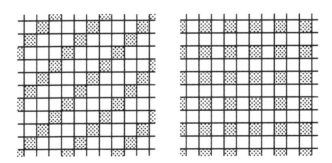

Figure 2

For each of the L-tetromino and the T-tetromino, since it contains the I-tromino, the minimum fraction is at most $\frac{1}{3}$.

To see that this is the best possible for the L-tetromino, consider a 2×3 rectangle. We must block off 2 of the 6 squares in order to bar the L-tetromino from the rectangle.

It is a bit more difficult to see that $\frac{1}{3}$ is the minimum fraction for the T-tetromino. Consider an infinite strip of width 3. If at least one square is blocked off in each column, there is nothing further to prove. So we assume that there is at least one blank column. This immediately means that the middle square in both adjacent columns must be blocked off. However, there are still two possible ways of placing a T-tetromino within these three columns. So two more squares must be blocked off.

If both of them are from the same column, this leads to the pattern in Figure 3 on the left. Otherwise, we will have the pattern in Figure 3 on the right. We may also have a mixture of both patterns. In any case, the minimum fraction is $\frac{1}{3}$.

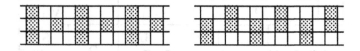

Figure 3

The chart below shows the minimum fractions for the pentominoes.

Pentominoes	F	I	L	N	P	T	U	V	W	X	Y	Z
Min. Frac.	$\frac{1}{4}$	$\frac{1}{5}$	$\frac{1}{4}$	$\frac{1}{4}$	$\frac{1}{4}$	$\frac{1}{4}$	$\frac{1}{4}$	$\frac{4}{13}$	$\frac{1}{4}$	$\frac{1}{5}$	$\frac{1}{4}$	$\frac{1}{3}$

Set D : Extremal and Mean Value Principles

This set of problems is also based on the Pigeonhole Principle. We approach it from a different direction, which is the more logical. We start with a fundamental result.

Extremal Value Principle.
In any finite non-empty collection of real numbers, there exist a maximum and a minimum.

Since the set is non-empty, we pick any of the real numbers. If it is the only number in the set, then it is both a maximum and a minimum. If there is another number, we can compare them. The larger one is the current maximum and the smaller one is the current minimum. The next number from the set is then compared with the current maximum and the current minimum. It may happen that one of them is replaced by the newcomer. Since the set is finite, this process must terminate. The numbers left standing as the current maximum and the current minimum are thus the maximum and the minimum.

It may happen that neither the maximum nor the minimum is unique. For instance, every number in the set may be the same. Then each is a maximum as well as a minimum.

Let us give an application of this principle. In a party, every boy dances with at least one girl, but no girl dances with every boy. Prove that there are two boys B_1 and B_2 along with two girls G_1 and G_2 such that B_1 has danced with G_1 but not G_2 while B_2 has danced with G_2 but not G_1.

Suppose we have somehow chosen B_1. Since B_1 dances with at least one girl, we have a choice for G_1. Since G_1 does not dance with every boy, we have a choice for B_2. Since B_2 dances with at least one girl, we have a choice for G_2 provided that she has not danced with B_1.

For the boys, let us count the number of girls with whom each has danced. This is a finite non-empty collection of real numbers, in fact, positive integers. By the Extremal Value Principle, there exists a minimum. So let B_1 be the boy who has danced with the least number of girls.

Among the girls who have danced with B_2, we seek one who has not danced with B_1. Suppose there are no such girls. Then B_1 has danced with every girl who has danced with B_2, along with G_1 who has not danced with B_2. Then B_1 has danced with more girls than B_2, contrary to the assumption that the number of girls who have danced with B_1 is minimum.

Mean Value Principle.
In any finite non-empty collection of real numbers, there exist one which is not above average and one which is not below average.

This follows easily from the Extremal Value Principle. The maximum cannot be below average, and the minimum cannot be above average. The astute readers may notice that one half of this result is what we have called the generalized Pigeonhole Principle. We have seen that the Pigeonhole Principle is a special case of this result. It has another half which states that *if there are less pigeons than holes, then there is at least one empty hole.*

Let us give an application of the Mean Value Principle. Consider a building with seven elevators, each stopping on three floors. These three floors do not have to be consecutive, and need not include the first floor. If we wish to go between any two floors, there is at least one elevator which stops on both, so that we never have to change elevators. What is the maximum number of floors in this convenient building?

With seven elevators each stopping on three floors, we have 21 elevator doors. These are the pigeons. The floors are the holes, and the pigeon representing an elevator door goes into the hole representing the floor on which the elevator door is located. Suppose we have eight or more floors. Then the average number of elevators per floor is less than three. By the Mean Value Principle, there is a floor on which at most two elevators stop. Since each of them can link that floor to only two other floors, we cannot reach every other floor without having to change elevators in some cases.

It follows that the convenient building can have at most seven floors. We now construct a plan for such a building with exactly seven floors, on each of which exactly three elevators stop.

We turn to a finite geometry with four points $(x, y) = (0,0)$, $(0,1)$, $(1,0)$ and $(1,1)$. We use arithmetic modulo 2 introduced in Set B. There are three pairs of parallel lines or lines with no common points.

Slopes	Lines	Points
0	$y = 0$	$(0,0)$, $(1,0)$
	$y = 1$	$(0,1)$, $(1,1)$
1	$y = x$	$(0,0)$, $(1,1)$
	$y = x + 1$	$(0,1)$, $(1,0)$
∞	$x = 0$	$(0,0)$, $(0,1)$
	$x = 1$	$(1,0)$, $(1,1)$

In Euclidean geometry, parallel lines do not meet, which explains why paintings in ancient time lack depth perception. This is corrected by the introduction of *projective geometry*, where parallel lines meet at infinity. We add an *ideal point* to every line, so that lines have the same ideal point if and only if they are parallel. Finally, an *ideal line* is added to link up all the ideal points. If we apply this procedure to the four-point geometry above, we have a structure called the *Fano plane*, shown in Figure 4.

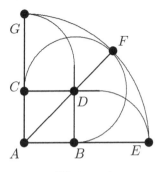

Figure 4

There are now seven points and seven lines. The ideal line is represented by a quarter of a circle, while the line $y = x + 1$ is represented by three quarters of a smaller circle. Parts of the lines $y = 1$ and $x = 1$ are curved. Here, every two lines determine a unique point, and every two points determine a unique line. This symmetry is an essential feature in projective geometry.

Returning to our convenient building, the seven elevators are the points A, B, C, D, E, F and G. The seven floors are the seven lines, ABE, CDE, ACG, BDG, ADF, BCF and EFG. An elevator stops on a floor if the point representing the elevator lies on the line representing the floor.

Problems 2 and 3 both involve colors. We give here a famous example. In a soccer club with six girls, every two exchange greeting cards. Each pair either exchange red cards or yellow cards. Prove that there are three girls who exchange cards of the same color among them.

Consider any player. She exchanges cards with five other girls. The average number of recipients per color is 2.5. Hence she must exchange cards of the same color with at least three others, say red. If any two of the three also exchange red cards, we have a red trio. Otherwise, the three of them form a yellow trio. The conclusion may be false if there are only five girls in the club. Let them sit at a round table. If they exchange red cards with their neighbors and yellow cards across the table, we have neither a red trio nor a yellow trio.

It turns out that if there are 18 girls instead, we will have either a red quad or a yellow quad, but 17 girls are not sufficient for that conclusion. A very general way of putting it comes from the social sciences. *If you have enough of it, something will happen.*

Joking aside, our example is a proto-type of a deep result called *Ramsey's Theorem*, a profound generalization of the Pigeonhole Principle.

In Problem 3, we try to paint the squares of the infinite chessboard with as few colors as possible so that when a specific polyomino is placed on the grid, it cannot cover two squares of the same color. For the monomino, 1 color is trivially sufficient. For the domino, we need 2 colors. The numbers of colors are 3 and 4 for the I-tromino and the V-tromino respectively. For the L-tetromino, the O-tetromino and the S-tetromino, 4 colors are sufficient.

Consider the region in Figure 5 with 5 unit squares. Every two of them may be covered by a suitable placement of the T-tetromino. This shows that 5 colors are necessary.

Figure 5

Figure 6 shows a 5-color infinite grid which establishes that 5 colors are sufficient for the T-tetromino.

4	1	5	2	3	4	1	5	2	3
2	3	4	1	5	2	3	4	1	5
1	5	2	3	4	1	5	2	3	4
3	4	1	5	2	3	4	1	5	2
5	2	3	4	1	5	2	3	4	1
4	1	5	2	3	4	1	5	2	3
2	3	4	1	5	2	3	4	1	5
1	5	2	3	4	1	5	2	3	4
3	4	1	5	2	3	4	1	5	2
5	2	3	4	1	5	2	3	4	1

Figure 6

112

Consider the region in Figure 7, with 12 unit squares. Every two of the 4 central squares may be covered by a suitable placement of the L-tetromino. Thus 4 colors, 1, 2, 3 and 4, are needed there. Moreover, any of these squares and any of the peripheral squares may be covered by a suitable placement of the L-tetromino. Thus these 4 colors may not be used again for the 8 peripheral squares. If only 7 colors are available, then one of the 3 additional colors, say 5, must be used on at least 3 peripheral squares. Paint any peripheral square in color 5. We may assume by symmetry that it is the one shown below. Then the 4 squares marked with crosses cannot be painted in color 5. Now 2 of the 3 blank squares must be in color 5, but any 2 of them may be covered by a suitable placement of the L-tetromino. This shows that 8 colors are necessary.

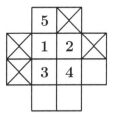

Figure 7

Figure 8 shows an 8-color infinite grid which establishes that 8 colors are sufficient for the L-tetromino.

1	2	5	6	1	2	5	6	1	2
3	4	7	8	3	4	7	8	3	4
5	6	1	2	5	6	1	2	5	6
7	8	3	4	7	8	3	4	7	8
1	2	5	6	1	2	5	6	1	2
3	4	7	8	3	4	7	8	3	4
5	6	1	2	5	6	1	2	5	6
7	8	3	4	7	8	3	4	7	8
1	2	5	6	1	2	5	6	1	2
3	4	7	8	3	4	7	8	3	4

Figure 8

113

The chart below shows the minimum number of colors for each of the pentominoes.

Pentominoes	F	I	L	N	P	T	U	V	W	X	Y	Z
Minimum #	8	5	8	8	8	8	8	9	6	5	8	9

Set E : Inequalities

There is one special case of the Pigeonhole Principle which has not yet been mentioned. What happens if the number of pigeons is equal to the number of holes? We may have two pigeons in the same hole, but then we will also have an empty hole. It may also happen that there is exactly one pigeon in each hole. In that case, we say there is a *one-to-one correspondence* between the pigeons and the holes.

Two little boys both boast that he has more marbles than the other. Neither of them can count. One way to settle this is for each to produce a marble at the time. This continues until one of them runs out of marbles. If the other also runs out at the same time, they have the same number of marbles. Otherwise, the other boy has more.

A statement about more or less is called an *inequality*, the counter part of an equation which is a statement of neither more nor less. Note that both the Extremal Value Principle and the Mean Value Principle are inequalities. Extremal value problems occur very often in mathematics. We have already encountered a few in Sets C and D. Problem 1 is another example.

To determine a maximum comes in three steps. In the first step, we claim that the answer is a certain value. In the second step, we have to show that it can be as large as that value. In the third step, we have to prove that it cannot be larger than that value. The last two steps together justify the claim made in the first step. Analogously, the same goes for determining a minimum. It should be mentioned that far too often, the third step is overlooked.

We give some more problems on inequalities.

1. We put 2018 dollars in several purses, and the purses in several pockets. The number of dollars in any pocket is less than the total number of purses. Is it necessarily true that the number of dollars in some purse is less than the total number of pockets?

2. The proportion of people with fair hair among people with blue eyes is more than the proportion of people with fair hair among all people. Which is greater, the proportion of people with blue eyes among people with fair hair, or the proportion of people with blue eyes among all people?

3. Ace has 3 dollars and 80 cents, Bea has an integral number of dollars and Cec has 60 cents. If any two of them have more money than the third, how much money does Bea have?

Here are the solutions.

1. Let m be the total number of pockets and n be the total number of purses. Let x be the maximum number of dollars in any pocket and y be the minimum number of dollars in any purse. Then we have $ny \leq 2018 \leq mx$. It follows that if $n > x$, then we must have $m > y$.

2. Let a, b, c and d denote the respective numbers of people with neither blue eyes nor fair hair, those with blue eyes but not fair hair, those with fair hair but not blue eyes, and those with blue eyes and fair hair. Then the proportion of people with fair hair among people with blue eyes is $\frac{c}{b+c}$, and the proportion of people with fair hair among all people is $\frac{c+d}{a+b+c+d}$. From the given condition, $\frac{c}{b+c} > \frac{c+d}{a+b+c+d}$, so that $\frac{c}{c+d} > \frac{b+c}{a+b+c+d}$. It follows that the proportion of people with blue eyes among people with fair hair is more than the proportion of people with blue eyes among all people.

3. Since Bea and Cec together have more money than Ace, Bea has more than 3 dollars and 20 cents. Since Ace and Cec together have more money than Bea, she has less than 4 dollars and 40 cents. Since Bea has an integral number of dollars, she has 4 dollars.

The last problem has a counterpart in geometry known as the *Triangle Inequality*: the total length of two sides of a triangle is always greater than the length of the third side.

Set F : Many-to-one Correspondences

The concept of one-to-one correspondence can be generalized to many-to-one correspondence, or even more general scenarios. In all three problems, we have a three-to-two correspondence. In each case, the consequence is that certain quantities must be multiples of $3 + 2 = 5$.

Problem 3 involves some unusual trading. We present here a more elaborate setting.

Pokémon cards have become a craze in Dilbertville. There are only three kinds of cards, and only three official dealers. Dogbert will trade one dog card for one cat card and one rat card, or vice versa. Catbert will trade one cat card for two dog cards and one rat card, or vice versa. Ratbert will trade one rat card for three dog cards and one cat card, or vice versa. No other trades are allowed.

Starting with one cat card, try to obtain

(a) some rat cards, but without any dog or cat cards.

(b) some dog cards, but without any cat or rat cards;

(c) at least two cat cards, but without any dog or rat cards.

For each task, either prove that it is impossible or find a sequence of trades to obtain the minimum number of cards of the specified type.

Here are the solutions.

(a) Clearly, we should conduct transactions with each trader in one direction only, getting more rat cards. Suppose we trade r times with Ratbert, d times with Dogbert and c times with Catbert. Counting dog cards, we have $3r - d + 2c = 0$. Counting cat cards, we have $1 - r + d - c = 0$. Addition yields $1 - 4r + c = 0$ or $c = 4r - 1$. Substituting back into either equation yields $d = 5r - 2$. Thus the smallest solution is $(r, d, c) = (1, 3, 3)$, resulting in only 1+3+3=7 rat cards. A possible sequence of trades is shown below:

Number	At the	After Trade with						
of Cards	Start	C	D	D	C	C	D	R
Rat	0	1	2	3	4	5	6	7
Dog	0	2	1	0	2	4	3	0
Cat	1	0	1	2	1	0	1	0

(b) Let s be the sum of the numbers of rat and cat cards. In any trade, the value of s changes by 0 or 2. Since $s = 1$ initially, we can never have only dog cards, because that means $s = 0$.

117

(c) This case is similar to (a) except that we must begin trading with Catbert in the wrong direction, and thereafter trading only to get more cat cards. We use the same notations as before. Counting rat cards, we have $1 - r + d - c = 0$. Counting dog cards, we have $2 + 3r - d - 2c = 0$. This time, we have $r = \frac{3(c-1)}{2}$ and $d = \frac{5(c-1)}{2}$. We must have $c > 1$ as otherwise we will end up with 1 cat card. Thus the smallest solution is $(x, y, z) = (3, 5, 3)$, resulting in only 3+5+3=11 cat cards. Any 11 trades with this distribution will work, after the initial one with Catbert.

Set G : Arithmetic Problems

This set consists of miscellaneous arithmetic problems. We present three more.

1. The price of 175 hotdogs is more than the price of 125 burgers but less than that of 126 burgers. If the price of each is an integral number of cents, can you buy three hotdogs and one burger with one dollar?

2. In the 40 tests Andrew had taken, he got 10 As, 10 Bs, 10 Cs and 10 Ds. A score is said to be *unexpected* if this particular score has appeared up to now fewer times than any of the other three scores. Without knowing the order of these 40 scores, is it possible to determine the number of unexpected ones?

3. A snail began crawling about a plane, starting from point O with constant speed, making a $60°$ turn every half hour. Prove that it can return to point O only after a whole number of hours.

Here are the solutions.

1. Let the price of a hotdog and a burger be h and b cents respectively, where h and b are integers. Then $125b < 175h < 126b$. Hence $7h > 5b$ and $25h < 18b$. It follows that $7h \geq 5b + 1$ and $25h \leq 18b - 1$. Multiplying the first inequality by 25 and the second by 7, we have $125b + 25 \leq 175h \leq 126b - 7$. Hence $b \geq 32$. From $7h \geq 5b + 1$, we have $7h \geq 161$ so that $h \geq 23$. Thus $3h + b \geq 69 + 32 = 101$. One dollar will not be enough.

2. Consider the first A, the first B, the first C and the first D that Andrew gets. The last one to come along must be unexpected, and none of the other three can be unexpected. The same applies to the second A, the second B, the second C and the second D that he gets, and so on. It follows that exactly 10 of the scores are unexpected.

3. Making $60°$ turns means that the snail is following a hexagonal net, moving from one vertex to an adjacent vertex in half an hour. The vertices can be marked so that exactly one of every pair of adjacent vertices is marked, as shown in Figure 9. Starting with the marked vertex O, the snail will always be at an unmarked vertex after an odd number of half hours have elapsed. Hence it must be on the hour when it returns to O.

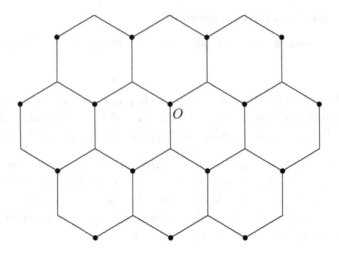

Figure 9

Set H : Divisibility Problems

Problem 1 deals with the concepts of *prime* numbers and *composite* numbers. A prime number is defined as a positive integer with exactly two positive divisors, namely, 1 and itself. Note that 1 is by definition not a prime number. A composite number is defined as a positive integer with more than two positive divisors. Thus 1 is neither prime nor composite. The first few prime numbers are 2, 3, 5 and 7, and the first few composite numbers are 4, 6, 8 and 9.

Interested readers may try to discover which composite numbers have exactly three positive divisors. There are infinitely many of them, and the class can be described in terms of the prime numbers.

We now present three problems on prime and composite numbers.

1. Can the ten digits 0, 1, 2, 3, 4, 5, 6, 7, 8 and 9 be arranged in a row so that no matter which six digits are removed, the remaining four digits, without changing their order, form a composite number?

2. Can the number $a + b + c + d$ be prime if a, b, c and d are positive integers such that $ab = cd$?

3. Do there exist

   (a) four;

   (b) five

   distinct positive integers such that the sum of any three of them is a prime number?

Here are the solutions.

1. Put 5 and the even digits in any order in the last six positions. If any of them remains, then the four-digit number is either divisible by 5 or by 2, and is hence composite. If all of them are removed, then the digits 1, 3, 7 and 9, which are in the first four positions in some order, must form a composite number. A simple way is to arrange them in the order 1, 3, 9 and 7, because the number 1397 is divisible by 11, and is hence composite.

2. Note that we have

$$a + b + c + d = \frac{a^2 + cd + ac + ad}{a} = \frac{(a+c)(a+d)}{a},$$

which is clearly composite since each factor in the numerator is larger than the denominator.

3. (a) The numbers 1, 3, 7 and 9 yield the primes 11, 13, 17 and 19.

   (b) Suppose there exist five such integers. Consider them in arithmetic modulo 3. If three of the five numbers belong to different classes, then their sum is a multiple of 3, which cannot be prime unless it is 3. However, this is impossible since the positive integers are distinct. If the five numbers are in at most two classes, then the Pigeonhole Principle guarantees that there will be three in the same class. Once again, their sum will be a multiple of 3. This is a contradiction.

Problems 2, 3 and 4 all deal with divisibility involving sums and products. We present three more such problems.

1. Do there exist 64 integers whose sum and product are both 64?

2. Find six distinct positive integers such that the product of any two of them is divisible by their sum.

3. Find ten different positive integers such that each is a divisor of their sum.

Here are the solutions.

1. Take one copy of 32, one copy of 2, m copies of 1 and n copies of -1. Since the number of integers is 64, we have $m+n = 62$. Since the sum of the integers is 64, we have $m - n = 30$. Hence $m = 46$ and $n = 16$. Since n is even, the product of the integers is $2 \times 32 = 64$.

2. Start with 1, 2, 3, 4, 5 and 6. The least common multiple of the 15 pairwise sums 1+2, 1+3, ..., 5+6 is $2^3 \times 3^2 \times 5 \times 7 \times 11 = 27720$. Multiplying it by 1, 2, 3, 4, 5 and 6 yields 27720, 55440, 83160, 110880, 138600 and 166320.

3. We claim that for any positive integer $n > 2$, there exist n distinct positive integers such that each is a divisor of their sum. For $n = 3$, we can take 1, 2 and 3. From a desirable set of n positive integers, we can obtain a desirable set of $n+1$ positive integers by adding the sum of the n numbers. This justifies the claim. In particular, if $n = 10$, we obtain 1, 2, 3, 6, 12, 24, 48, 96, 192 and 384.

Set I : Digital Problems

Problems 1 and 2 are both about tempering with the internal structure of multi-digit numbers. We give another example here.

Two 2009-digit numbers are such that it is possible to delete 9 digits from each of them to obtain the same 2000-digit number. Prove that it is also possible to insert 9 digits into the given numbers so as to obtain the same 2018-digit number.

This problem can be solved as follows. To the 2000-digit number, add back the 9 digits deleted from the two 2009-digit numbers. Digits from different 2009-digit numbers inserted between the same two digits of the 2000-digit number can be arranged in any order. This yields a 2018-digit number. It is obtainable from either 2009-digit number by inserting the 9 digits from the other 2009-digit number.

Problem 3 mixes digits with age. We give another example.

If the two digits of the father's age are reversed, we get the son's age. Tomorrow, the father will be twice as old as the son. How old is the son today?

Let x and y be the digits of the father's age which is $10x + y$. The son's age is $10y + x$, and $y < x$. Most people will proceed as follows. From $10x + y = 2(10y + x)$, we have $8x - 19y = 0$. The smallest positive integral value for x is 19, but this contradicts the condition that x is a digit. Hence the problem has no solutions.

However, in jumping to this conclusion, we have overlooked the possibility that the next day may be the birthday of either the father or the son, or both. Thus there are three other cases to consider.

Case 1. The next day will be the birthday of both the father and the son. Then $10x + y + 1 = 2(10y + x + 1)$, which simplifies to $8x - 19y = 1$. By inspection, the smallest positive integral value for x is 12. Thus this case is also impossible.

Case 2. The next day will be the birthday of the father but not the son. Then $10x + y + 1 = 2(10y + x)$, which simplifies to $8x - 19y = -1$. By inspection, we have $x = 7$ and $y = 3$. So the father is 73 and the son is 37.

Case 3. The next day will be the birthday of the son but not the father. Then $10x + y = 2(10y + x + 1)$, which simplifies to $8x - 19y = 2$. By inspection, we have $x = 5$ and $y = 2$. So the father is 52 and the son is 25.

In summary, the son's age is either 25 or 37.

Digital problems come in a great variety. A well-known example is to find a number which becomes 4 times as large when its last digit is moved to the front.

We may approach the problem this way. The digit moved is at least 4. Suppose it is 4. We perform the following divisions. We simply add the preceding quotient to the dividend, until we arrive at an exact division with 4 as the last digit of the quotient.

$$\begin{array}{cccccc} 1 & 10 & 102 & 1025 & 10256 & 102564 \\ \overline{4)4} & \overline{4)41} & \overline{4)410} & \overline{4)4102} & \overline{4)41025} & \overline{4)410256} \end{array}$$

We have $102465 \times 4 = 410256$. If the first digit is 5, 6, 7, 8 or 9, we get other answers: $4 \times 128205 = 512820$, $4 \times 158346 = 615834$, $4 \times 179487 = 717948$, $4 \times 205128 = 820512$ and $4 \times 230769 = 923076$. We will meet one of these six-digit numbers in Set L.

The Japanese genius **Nobiyuki Yoshigahara** was arguably the most inventive as well as the most whimsical puzzle creator the world had ever seen. The following astounding digital wonder came from Nob, as he was fondly known. He presented a pair of numbers, 286794 and 5103. Apart from the fact that between the two of them, they use all ten digits, there is nothing really remarkable. Then he permuted the digits of each number to obtain another pair, 479682 and 3051. So what is so astounding so far? The secret will be revealed when the product of the two pair of numbers are compared.

Nob also asked for a number which becomes 6 times as large when its last digit is moved to the front.

Set J : Tests of divisibility

All three problems deal with divisibility properties of numbers based on the digits in their base-ten expressions. Analyzing them leads to the tests of divisibility, procedures for determining whether a large positive integer is divisible by a small positive integer, without performing the actual division.

Every positive integer is divisible by 1, and a positive integer is divisible by the base number 10 if and only if its last digit is 0. We seek tests of divisibility by positive integers up to 12.

Among them, 2 and 5 are divisors of 10. Hence a positive integer is divisible by 2 if and only if its last digit is 0, 2, 4, 6 or 8, and divisible by 5 if and only if its last digit is 0 or 5.

Divisibility tests for 4 and 8 follow similar patterns, in that $4 = 2^2$ is a divisor of $10^2 = 100$, and $8 = 2^3$ is a divisor of $10^3 = 1000$. Suppose we wish to determine whether 194711276 is divisible by 4 or 8. First we write $194711276 = 1947112 \times 100 + 76$. The first term is always divisible by 4 since it is a multiple of 100. Since 76 is a multiple of 4, so is 194711076. Now $194711276 = 194711 \times 1000 + 276$. Since 276 is not divisible by 8, neither is 194711276. In summary, a number is divisible by 4 if and only if the number formed of its last two digits is divisible by 4, and divisible by 8 if the number formed of its last three digits is divisible by 8.

Next, we turn to 9, which is 1 less than the base number 10, and 3, which is a divisor of 9. Note that $9 = 10 - 1$, $99 = 100 - 1$, $999 = 1000 - 1$, $9999 = 10000 - 1$, ..., are all multiples of 9. Take the number 19473. We can write it as follows.

$$
\begin{aligned}
19473 &= 1 \times 10000 + 9 \times 1000 + 4 \times 100 + 7 \times 10 + 3 \\
&= 1 \times (9999+1) + 9 \times (999+1) + 4 \times (99+1) + 7 \times (9+1) + 3 \\
&= 9 \times (1 \times 1111 + 9 \times 111 + 4 \times 11 + 7) + (1 + 9 + 4 + 7 + 3).
\end{aligned}
$$

The first term is divisible by 9. Since $1 + 9 + 4 + 7 + 3 = 24$ is not divisible by 9, neither is 19473. However, 19473 is divisible by 3 since so is 24. This is because the first term, as a multiple of 9, is automatically a multiple of 3.

In summary, a number is divisible by 9 if and only if the sum of its digits is divisible by 9, and divisible by 3 if and only if the sum of its digits is divisible by 3.

The method of *casting out 9s* is a shortcut to checking if the sum of the digits of a number is divisible by 9. Whenever we see a digit 9, we can strike it out. Whenever we see two digits with sum 9, namely, 1 and 8, 2 and 7, 3 and 6 or 4 and 5, we can strike them out too.

We now turn to 11, which is 1 more than the base number 10. Note that $11 = 10 + 1$, $99 = 100 - 1$, $1001 = 1000 + 1$, $9999 = 10000 - 1$, ..., are all multiples of 11. Again, we take the number 19473.

$$\begin{aligned} 19473 \ &= \ 1 \times 10000 + 9 \times 1000 + 4 \times 100 + 7 \times 10 + 6 \\ &= \ 1 \times (9999+1) + 9 \times (1001-1) + 4 \times (99+1) + 7 \times (11-1) + 6 \\ &= \ 11 \times (1 \times 909 + 9 \times 91 + 4 \times 9 + 7) + (1 - 9 + 4 - 7 + 3). \end{aligned}$$

The first term is divisible by 11. Since $1 - 9 + 4 - 7 + 3 = -8$ is not divisible by 11, neither is 19473. In summary, a number is divisible by 9 if and only if the sum of its digits, taken with alternating signs, is divisible by 11.

The number 3 has no common divisor greater than 1 with either 2 or 4. Hence a number is divisible by $6 = 2 \times 3$ if and only if it is divisible by 2 and by 3. Similarly, a number is divisible by $12 = 3 \times 4$ if and only if it is divisible by 3 and 4. However, a number divisible by 2 and 6 is not necessarily divisible by $2 \times 6 = 12$. A counterexample is the number 6. The failure of the test is because 2 and 6 have a common divisor greater than 1, namely, 2.

This leaves only the number 7 among the first dozen. There are so-called tests of divisibility for 7, but they are not really much shorter than performing the actual division.

Set K : Squares and Square Roots

There is actually a second solution to Problem 1, in that

$$351364183^2 = 123456789095257489.$$

A method or algorithm for finding square roots, akin to long division, was presented in *Liber Abaci* (The Book of Calculations), written in medieval time by Leonardo Pisano (commonly known as Fibonacci). We demonstrate with the number 20187049. We divide the digits of the number into pairs, 20|18|70|49. In the case where the number has an odd number of digits, the first block has only one digit.

1. The first block is 20. Since $4 \times 4 < 20 < 5 \times 5$, the first digit of the "quotient" is 4. The first remainder is 4.

2. Add 100 times the first remainder to the next block and obtain 418. Multiply the quotient so far by 20 to obtain 80. The second digit of the quotient is 5 since $84 \times 4 < 418 < 85 \times 5$. The second remainder is 82.

3. Add 100 times the second remainder to the next block and obtain 8270. Multiply the quotient so far by 20 and obtain 880. The third digit of the quotient is 9 since $889 \times 9 < 8270$. The third remainder is 269.

4. Add 100 times the third remainder to the next block and obtain 26949. Multiply the quotient so far by 20 and obtain 8980. The four digit of the quotient is 3 since $8893 \times 3 = 26949$, and there is no remainder.

		4	4	9	3
		20	18	70	49
4	16				
80	4	18			
84	3	36			
880	82	70			
889	80	01			
8980	2	69	49		
8983	2	69	49		

We use a little algebra to help us understand why this algorithm works. Let $10a + b$ be a two-digit number. Then

$$(10a + b)^2 = 100a^2 + 20ab + b^2 = 100a^2 + b(20a + b).$$

The coefficient 100 for a^2 is the reason why we divide the number into blocks of two digits.

Suppose we wish to find the integer part of the square root of 2018. The digit a satisfies $a^2 \leq 20 < (a+1)^2$, and is easily seen to be 4. Subtracting 1600 from 2018, we have 418. Now $20a = 80$, and we want the digit b such that $b(80 + b) < 418 < (b+1)(81+b)$. It is easy to see that $b = 4$. Indeed $44^2 < 2018 < 45^2$.

Let $100a + 10b + c$ be a three-digit number. We have

$$(100a + 10b + c)^2 = 10000a^2 + 2000ab + 100b^2 + 200ac + 20bc + c^2.$$

Suppose we wish to find the integer part of the square root of 201870. From $a^2 \leq 20 < (a+1)^2$, we have $a = 4$. Subtracting 160000 from 201870, we have
$$41870 = 100b(20a + b) + c(20(10a + b) + c).$$

The digit $b = 4$ satisfies $b(80+b) \leq 418 < (b+1)(81+b)$. Subtracting 33600 from 41870, we have $c(880 + c) < 8270 < (c+1)(881+c)$. It is easy to see that $c = 9$. Indeed $449^2 < 201870 < 450^2$.

It is not difficult to generalize to numbers with more than three digits. It should now be clear how $\sqrt{20187049} = 8893$ is obtained. We give below the computation showing $\sqrt{123456789095257489} = 351364183$.

		3	5	1	3	6	4	1	8	3
		12	34	56	78	90	95	25	74	89
3	9									
60	3	34								
65	3	25								
700	9	56								
701	7	01								
7020	2	55	78							
7023	2	10	69							
70260	45	09	90							
70266	42	15	96							
702720	2	93	94	95						
702724	2	81	08	96						
7027280	12	85	99	25						
7027281	7	02	72	81						
70272820	5	83	26	44	74					
70272828	5	62	18	26	24					
702728360	21	08	18	50	89					
702728363	21	08	18	50	89					

Problems 2 and 3 both involved modular arithmetic, introduced in Set B, and tests of divisibility, introduced in Set J.

In arithmetic modulo 3, every integer is represented by one of 0, 1 or 2. Since $0^2 = 0$, $1^2 = 1$ and $2^2 = 4 \equiv 1 \pmod 3$, squares are never represented by 2. Similarly, squares are never represented by 2 or 3 in arithmetic modulo 4. The readers are invited to explore the situation with other positive integers as the modulus.

Set L : Cyclic Numbers

All three problems deal with six-digit numbers. The most interesting and well-known six-digit number is 142857. We have seen in Set C that it is one cycle of the repeating decimal expansion of $\frac{1}{7}$.

Note that $14 + 28 + 57 = 99$ and $142 + 857 = 999$. Moreover, it has the following nice properties.

$$
\begin{array}{rcl}
1 \times 142857 &=& 142857, \\
2 \times 142857 &=& 285714, \\
3 \times 142857 &=& 428571, \\
4 \times 142857 &=& 571428, \\
5 \times 142857 &=& 714285, \\
6 \times 142857 &=& 857142, \\
7 \times 142857 &=& 999999.
\end{array}
$$

What happens if we multiply 142857 by larger numbers? For instance, $142857^2 = 20408122449$. Dividing the product into blocks of length 6 from the right and adding them up, we have $20408 + 122449 = 142857$. If the sum has more than 6 digits, the process is repeated. If the multiplier is 1 more than a multiple of 6, then 999999 will emerge.

Martin Gardner presented the following magic trick based on this cyclic number. Prepare a paper ring with the digits 1, 4, 2, 8, 5 and 7 evenly distributed. Make it flat so that 1, 4 and 2 are on one side and 8, 5 and 7 are on the other. Put the flattened paper ring inside an envelope and seal it. Force the number 142857 on the audience in some manner, and ask the audience to roll a standard six-sided die. Then declare that the product of 142857 and the die-roll has been predicted.

If the die-roll is 1, cut the envelope at A in Figure 10, so that the paper ring opens to reveal 142857. If the die-roll is 3 or 5, cut the envelope at B. Pass the scissor through and above the paper ring to reveal 428571 or through and below the paper ring to reveal 714285. Apply analogous procedures if the die-roll is 2 or 4 (cut at C) or 6 (cut at D).

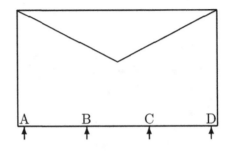

Figure 10

Another interesting six-digit number is 076923. Just as before, we have $07 + 69 + 23 = 99$ and $076 + 923 = 999$. This time, we have the following.

$$
\begin{aligned}
1 &\times 076923 = 076923, \\
3 &\times 076923 = 230769, \\
4 &\times 076923 = 307692, \\
9 &\times 076923 = 692307, \\
10 &\times 076923 = 769230, \\
12 &\times 076923 = 923076, \\
13 &\times 076923 = 999999.
\end{aligned}
$$

The last line suggests that 076923 is one cycle of the repeating decimal expansion of $\frac{1}{13}$. We can verify that indeed it is. However, why do we keep 1, 3, 4, 9, 10 and 12, and skip over 2, 5, 6, 7, 8 and 11? Let us see what happens with those skipped numbers as the multiplier.

$$
\begin{aligned}
2 &\times 076923 = 153846, \\
5 &\times 076923 = 384615, \\
6 &\times 076923 = 461538, \\
7 &\times 076923 = 538461, \\
8 &\times 076923 = 615384, \\
11 &\times 076923 = 846153.
\end{aligned}
$$

So 076923 has a companion number 153846. It also has the property that $15 + 38 + 46 = 99$ and $153 + 846 = 999$. In arithmetic modulo 13, we have $1^2 = 1$, $2^2 = 4$, $3^2 = 9$, $4^2 = 3$, $5^2 = 12$ and $6^2 = 10$. Thus the first set of multipliers are the squares, and the second set are the non-squares.

In arithmetic modulo 7, the squares are $1^2 = 1$, $2^2 = 4$ and $3^2 = 2$. The non-squares are 3, 5 and 6. We can see the following pattern:

$$
\begin{aligned}
1 &\times 142857 = (14)(28)(57), \\
2 &\times 142857 = (28)(57)(14), \\
4 &\times 142857 = (57)(14)(28),
\end{aligned}
$$

$$
\begin{aligned}
3 &\times 142857 = (42)(85)(71), \\
5 &\times 142857 = (71)(42)(85), \\
6 &\times 142857 = (85)(71)(42).
\end{aligned}
$$

Set M : Problems on Money

M is for money. The modern world is obsessed with it, which makes it not as nice a place as it used to be. Even in mathematics, problems involving money occur very frequently, such as the three in this set. Fortunately, we are only obsessed with mathematics, not with money. **M** is for mathematics!

That being said, we will present three more problems involving money.

1. Betty, Hetty and Letty went on a picnic. Betty brought five sandwiches and Hetty brought four. Letty forgot to bring any. After sharing the sandwiches equally, Letty paid the other two nine dollars. How many dollars should go to Betty?

2. Betty, Hetty and Letty spent the night in a hostel. The clerk charged each of them ten dollars. Later a bell-hop came, declaring that the clerk had forgotten about a 10% promotional discount, and would refund them three dollars. However, the bell-hop had only a five-dollar bill. So the girls gave him two dollars in change. Now the girls had paid thirty dollars initially. The twenty-seven dollars they were actually paying, plus the two dollars they gave to the bell-hop, only added up to twenty-nine dollars. What happened to the missing dollar?

3. The Gift Shop of the hostel had a beautiful doll which the girls liked very much. The cost was an integral number of dollars less than one hundred, but still more than any of them could afford. Each of them also had an integral number of dollars. Betty said to Hetty, "If you lend me one third of your money, I could just pay for the doll." Hetty said to Letty, "If you lend me one fourth of your money, I could just pay for the doll." Letty said to Betty, "If you lend me one fifth of your money, I could just pay for the doll." How much did the doll cost?

Here are the solutions.

1. Many people conclude that Betty should get five dollars and Hetty four, proportional to the numbers of sandwiches they provided. However, this is not correct. Since nine dollars were paid for a fair share of three sandwiches, each sandwich is worth three dollars. Betty provided five and ate three. So she should get six dollars while the other three dollars go to Hetty.

2. There was no missing dollar. The two dollars which the bell-hop received should not have been added to the twenty-seven dollars the girls actually paid. Instead, it should be added to the thirty dollars they had paid initially, for a total of thirty-two dollars. In return, the girls got twenty-seven dollars worth of accommodation and five dollars from the bell-hop.

3. Suppose Betty had $5x$ dollars, Hetty had $3y$ dollars, Letty had $4z$ dollars, and the doll cost w dollars. Then $5x+y = 3y+z = 4z+x = w$. Now $15x + 3y = 3w$. Subtracting from this $3y + z = w$, we have $15x - z = 2w$ so that $60x - 4z = 8w$. Adding to this $4z + x = w$, we have $61x = 9w$. Since 61 and 9 have no common divisor greater than 1 and $w \leq 100$, we must have $w = 61$. Incidentally, $x = 9$, $y = 16$ and $z = 13$, so that Betty had 45 dollars, Hetty had 48 dollars and Letty had 52 dollars.

Set N : Magic Configurations

Problem 1 asks for the construction of a magic rectangle, using consecutive positive integers starting from 1. This is a variation of the classic problem of constructing magic squares, a topic with a long history spanning over many cultures.

The sum of the numbers in each row and column of a magic square is called its *magic constant*. The side length of a magic square is called its *order*. A magic square of order 1 is trivial, and it is easy to see that a magic square of order 2 does not exist. We now give a construction of a magic square of order 3.

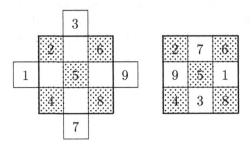

Figure 11

In Figure 11 on the left, we write down the numbers 1 to 9 in order in three diagonals. The four numbers outside the 3 × 3 table move straight inwards by three spaces. The resulting magic square is shown in Figure 11 on the right. The magic constant is 15.

This method generalizes to the construction of magic squares of odd order. Figure 12 illustrates the case of order 5.

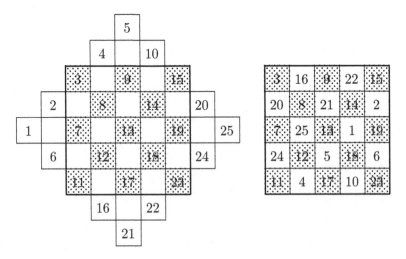

Figure 12

134

The magic square in Figure 11 on the right is the unique order 3 magic square. It appears frequently in mathematical folklore. In ancient China, it was called the *lo-shu*. It has the additional property that the sum of the three numbers on each long diagonal is also 15. Based on this magic square, the great **Martin Gardner** introduced a delightful game called *The Game of Fifteen*. We leave the joy of discovering its connection with *lo-shu* to the readers.

There are nine cards on the table, numbered 1, 2, 3, 4, 5, 6, 7, 8 and 9 respectively. Anna and Boris take turns picking up a card from the table, Anna going first. Whichever player first obtains three cards with sum exactly 15 wins the game. There may be additional cards in the hand of the winning player. If neither player has won when all nine cards have been picked up, the game is a draw. Does either Anna or Boris have a winning strategy?

Problem 2 takes us to a 4×4 square with some magic properties, but not exactly a magic square. Beyond order 3, the complexity of magic squares rises. Moreover, the construction of magic squares of even order is harder. Figure 13 shows one of 880 distinct magic squares of order 4. It appears in Albrecht Dürer's famous engraving *Melancolia* in the year 1514.

16	3	2	13
5	10	11	8
9	6	7	12
4	**15**	**14**	1

Figure 13

Problem 3 takes us to a magic configuration in space. The analog to the concept of magic squares is naturally the concept of magic cubes. The history is comparatively short, but the amount of interesting material is still astonishing. Once again, a magic cube of order 1 is trivial and a magic cube of order 2 does not exist. Figure 14 shows one of four distinct magic cube of order 3. The sum of the three numbers in each row, column or stack is 42, as is that in any diagonal passing through the central cube. In *The Hitchhiker's Guide to the Galaxy*, 42 is the answer to the universe!

10	26	6
24	1	17
8	15	19

23	3	16
7	14	21
12	25	5

9	13	20
11	27	4
22	2	18

Figure 14

Set O : Logic Problems

Problems involving liars and truth-tellers as in Problem 1 are very old. Here is a more elaborate one.

Each of Austin, Dustin and Justin is either a liar or a truth-teller. A stranger asks Austin, "Are you a liar or a truth-teller?" Austin's answer is so indistinct that the stranger cannot make out what he says. The stranger than asks Dustin, "What does Austin say?" Dustin replies, "Austin says that he is a liar." Justin jumps in, "Don't believe Dustin; he is lying!" What are Dustin and Justin?

When the great logician **Raymond Smullyan** came upon this problem, it immediately struck him that Justin did not really function in any essential way; he was sort of an appendage. That is to say, the moment Dustin speaks, one can tell without Justin's testimony that Dustin is lying. This is because whether Austin is a liar or a truth-teller, he can never say that he is a liar. Smullyan eliminates this blemish by modifying the problem as follows.

Suppose the stranger, instead of asking Austin what he is, asks Austin, "How many truth-tellers are among you?" Again Austin answers indistinctly. So the stranger asks Dustin, "What does Austin say?" Dustin replies, "Austin says that there is one truth-teller among us." Then Justin says, "Don't believe Dustin; he is lying!" Now what are Dustin and Justin?

Smullyan calls the truth-tellers Knights and the liars Knaves. They are introduced in his outstanding *What Is the Name of This Book?* and these have become standard terms. Amidst intriguing problems in fairytale settings, this book has a hidden surprise. If one works diligently and intelligently through the relevant puzzles, one would have a proof of a deep result in meta-mathematics called Gödel's Theorem.

Smullyan has written more than a dozen fascinating books. A few choice titles are *Alice in Puzzleland, The Riddle of Scheherazade, The Chess Mysteries of Sherlock Holmes* and *The Chess Mysteries of the Arabian Knights.* Apart from his profound research and prolific book writing, Raymond delighted in giving lectures to the public and especially to school children.

He would begin by asking the audience, "I have good news and bad news. Which do you want first?" Suppose they said, "Bad news." He said, "The bad news is that there is no good news." Outraged, the audience demanded, "What is the good news then?" "The good news is that the bad news is not true!"

Baron Münchhausen, featured in Problems 2 and 3, is a famous character in the folklore of mathematics and beyond. He is famous for telling tall tales which may somehow turn out to be true.

Once he said he tied his horse to a cross on a snowfield, and went to sleep beside it. When he woke up the next day, he found himself in the middle of a busy market. His horse was nowhere to be seen, until he heard it neighing high above his head. It was tied to the steeple on top of a nearby church! The explanation was that the snow was so heavy the night before that the town was covered up except for the steeple, and then the snow melted away.

Set P : Coin Weighing Problems

Problems 1 and 2 deal with the partial ranking of a set of coins with distinct weights. In *The Puzzling Adventures of Dr. Ecco* by the prolific problem proposer **Dennis Shasha**, one of the problems asks for the complete ranking of eight coins with distinct weights. Moreover, four balances can be used simultaneously, and we wish to use as few rounds of weighings as possible.

Let us first downsize the problem to two coins. Clearly, a single weighing is both necessary and sufficient. So the task requires just one round. If we have more coins, we use the divide and conquer approach. For four coins, we can obtain two sorted pairs (A_1,A_2) and (B_1,B_2) in the first round, with A_1 and B_1 the heavier in the respective pairs. From here, we can proceed in two ways.

Most people would weigh A_1 against B_1 and A_2 against B_2 in the second round. This will determine first and last place. If the remaining two coins happen to have been weighed against each other in the first round, the ranking is already complete. Otherwise, a third round will decide second and third place. This method is known as the *odd-even merge-sort* since the coins with odd indices are put into one group while those with even indices are put into another group.

We could also weigh A_1 against B_2 and B_1 against A_2 in the second round. If A_1 and B_1 are heavier, they must be the top two while A_2 and B_2 are the bottom two. The ranking can be completed in the third round. On the other hand, if either A_1 or B_1 is lighter in the second round, we already have a complete ranking. This method is known as the *upside-down merge-sort*, and the reason is obvious.

With eight coins, we can obtain two sorted quartets (A_1,A_2,A_3,A_4) and (B_1,B_2,B_3,B_4) in three rounds. Using the odd-even merge-sort, we merge (A_1,A_3) and (B_1,B_3) into a sorted quartet (C_1,C_2,C_3,C_4), and (A_2,A_4) and (B_2,B_4) into a sorted quartet (D_1,D_2,D_3,D_4). Clearly, C_1 is the heaviest coin while D_4 is the lightest. It can be shown that the second and third places belong to C_2 and D_1, the fourth and fifth places to C_3 and D_2, and the sixth and seventh places to C_4 and D_3. A sixth round will complete the ranking. Using the upside-down merge-sort, we weigh A_1 against B_4, A_2 against B_3, A_3 against B_2 and A_4 against B_1. If either A_1 or B_1 is the lighter coin in the respective pair, the ranking is complete. Hence we assume both are heavier. If both A_2 and B_2 are the heavier coins, then (A_1,A_2) and (B_1,B_2) are the top four, while (A_3,A_4) and (B_3,B_4) are the bottom four. The ranking can be completed in two more rounds. Suppose A_3 is heavier than B_2. Then the top four are (A_1,A_2,A_3) and B_1 while the bottom four are A_4 and (B_2,B_3,B_4). In the fifth round, we weigh A_2 against B_1 and A_4 against B_3. The ranking can be completed in the sixth round.

The readers are invited to explore the possibility of completing the ranking of eight coins in five rounds.

Problems 3 and 4 involve the identification of counterfeit coins. An effective method makes use to *ternary codes*, that is, strings of 1s, 2s and 3s. The choice of three is natural, because there are three possible outcomes in a weighing. We may have equilibrium, the left pan may be heavier, or the right pan instead.

A classic problem is to identify one heavier counterfeit coin among nine coins, the other eight being genuine and have equal weights. Since we have two weighings, we use ternary codes of length 2, of which there are $3^2 = 9$. Assign the codes 11, 12, 13, 21, 22, 23, 31, 32 and 33 to the nine coins. In the first weighing, put all the coins with 1 as the first digit in their codes on the left pan and those with 3 as the first digit in their codes on the right pan. In the second weighing, we use the second digit of the codes instead.

In the first weighing, if we have equilibrium, we know that the first digit in the code of the counterfeit coin is 2. If the left pan is heavier, it is 1. If the right pan is heavier, it is 3. Similarly, the second weighing will determine the second digit of the code, and the counterfeit coin is identified.

A much more difficult problem involves twelve coins, one of which is counterfeit, and we do not know in advance if it is heavier or lighter than a genuine coin. In three weighings, we wish to identify the counterfeit coin, and also determine whether it is heavier or lighter.

Since we have three weighings, we use ternary codes of length 3. We have a problem right from the start, because there are $3^3 = 27$ ternary codes of length 3, but we only have twelve coins. We will eliminate three codes, and assign two of the remaining codes to each coin, one in the event that the counterfeit coin is heavy, and other in the event that the counterfeit coin is light.

It is perhaps not surprising that we would eliminate 111, 222 and 333. The remaining codes form complementary pairs in that the sum of the corresponding digits in the two codes is always 4. Then one of the codes is assigned as primary while the other as secondary. Such an assignment is shown in the chart below.

Coins	A	B	C	D	E	F
Primary Codes	112	331	121	122	123	313
Secondary Codes	332	113	323	322	321	131
Coins	G	H	I	J	K	L
Primary Codes	312	311	233	232	231	223
Secondary Codes	132	133	211	212	213	221

139

Apart from forming complementary pairs, there is another requirement in the assignment of codes. For each of the first, second and third digit in the primary codes, there must be exactly four 1s, four 2s and four 3s. Then the secondary code will automatically have the same property.

In each weighing, we put in the left pan coins with 1 as the relevant digit of the primary codes, and in the right pan coins with 3 as the relevant digit of the primary codes.

Weighings	Left Pan	Right Pan
First	A,C,D,E	B,F,G,H
Second	A,F,G,H,	B,I,J,K
Third	B,C,H,K	E,F,I,L

Since each coin is involved in at least one weighing, we cannot have equilibrium all three times. This is why we eliminate the ternary code 222. Since each of coins B, E, F, G, H and K appears in different pans in different weighings, we cannot have left heavy all three times or right heavy all three times either. Thus 111 and 333 are also eliminated.

Suppose the outcomes of the three weighings are left heavy, left heavy and right heavy. On the assumption that the counterfeit coin is heavy, its primary code should be 113. Now 113 is the secondary code of coin B. Therefore, coin B is counterfeit and is lighter than a genuine coin.

Set Q : Geometric Configurations

Problem 1 essentially asks for a configuration of points in the plane so that each is equidistant from three other points. Our solution involves 16 points. Find a configuration involving a smaller number of points. Note however that here, we cannot replace the points by tangent circles. Find also a configuration in which each point is equidistant from four other points.

Problem 2 involves a chessboard configuration. Here is another one. Six rooks are placed on the squares of a 6×6 chessboard so that no two of them attack each other. So each vacant square is under attack from exactly two rooks. Is it possible that for each vacant square, the two attacking rooks are at the same distance away? Is it possible that for each vacant square, the two attacking rooks are at different distances away?

Problem 3 involves numbers in a geometric configuration. Here we change the setting to a castle with five intersecting walls in the shape of a regular five-pointed star. There are ten watchtowers at the junctions of two walls. Each guard stays in a watchtower and watches over the entire lengths of both walls meeting there. What is the minimum number of guards required so that each wall is watched over by at least five guards?

Incidentally, the King is not happy with the design of the castle. He objects to the fact that all ten watchtowers are subject to direct attack from outside. He wants a re-design, still with five walls intersecting pairwise at ten watchtowers. However, there should be one watchtower protected by the walls, which the King can use as his Royal residence. Can his desire be met?

We now give the solutions to the new problems.

In the first problem, note that each vertex of a square is equidistant from two other vertices. We use this idea to solve both parts of the problem. In Figure 15 on the left, we use two interlocking squares, and on the right, three interlocking squares.

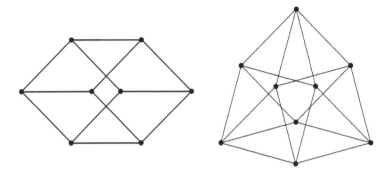

Figure 15

In the second problem, if we place the six rooks on the same diagonal, then every vacant square is attacked by two rooks from the same distance away. On the other hand, if the rooks are placed as non-attacking queens, as shown in Figure 16, then every vacant square is attacked by two rooks from different distances away.

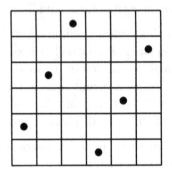

Figure 16

In the third problem, we have 5 walls each requiring 5 guards. Since each guard watches over two walls, the number of guards cannot be less than 13 since $(5 \times 5) \div 2 = 12.5$. We can get by with 13 guards by placing them all on watchtowers at the vertices of the central pentagon. Put 2 guards each on two non-adjacent watchtowers, and 3 guards each on the other three watchtowers. Then each wall is watched over by at least 5 guards.

After some brainstorming, the Royal Engineers come up with the design shown in Figure 17. Now the Queen wants her own protected watchtower too, and the Royal Engineers have to start all over again.

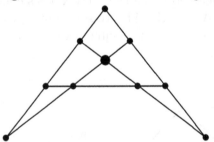

Figure 17

Set R : Problems on Coloring

Colors are involved in the statements of Problems 1 and 3 but not in the statement of Problem 2. Instead, coloring is used as a technique for solving Problem 2. We present three more problems.

1. There is at least one boy and at least one girl among twenty children in a circle. None of them is wearing more than one T-shirt. For each boy, the next child in the clockwise direction is wearing a blue T-shirt. For each girl, the next child in the counterclockwise direction is wearing a red T-shirt. Is it possible to determine the exact number of boys in the circle?

2. Numbers from 1 to 1000 are arranged arbitrarily around a circle. Prove that it is possible to form 500 non-intersecting line segments, each joining two of the numbers whose absolute difference is at most 749.

3. Each vertex of a regular 45-gon is red, yellow or green, and there are 15 vertices of each color. Prove that we can choose three vertices of each color so that the three triangles formed by the chosen vertices of the same color are congruent to one another.

Here are the solutions.

1. We claim that the boys and girls must alternate along the circle. Suppose to the contrary that two girls are next to each other. Then for some boy, the two children clockwise from him are both girls. The first girl must be wearing a blue T-shirt because of the boy, and a red T-shirt because of the other girl. This is a contradiction. Similarly, we cannot have two boys next to each other, and our claim is justified. It follows that the number of boys must be 10. In fact, all the boys are wearing red T-shirts and all the girls are wearing blue T-shirts.

2. Paint the points numbered 251 to 751 inclusive red and the others blue. Join each red point to a distinct blue point by a line segment. Among all possible configurations of the 500 segments, choose one in which the total length of the segments is minimum. Suppose two segments RB and $R'B'$ intersect at some point P. We replace them by RB' and $R'B$. By the Triangle Inequality,

$$RB + R'B' = RP + BP + R'P + B'P > RB' + R'B.$$

However this contradicts our minimality assumption. It follows that no two of the 500 segments in the minimal configuration intersect. Hence $1000 - 251 = 749 = 750 - 1$ is the maximum difference between a red and a blue point.

3. Copy the regular 45-gon onto a piece of transparency and mark on it the 15 red points. Call this the Red position, and rotate the piece of transparency about the centre of the 45-gon $8°$ at a time. For each of the 45 positions, count the number of matches of yellow points with the 15 marked points. Since each of the 15 yellow points may match up with any of the 15 marked points, the total number of matches is $15 \times 15 = 225$, so that the average number of matches per position is 5. However, in the Red position, the number of matches is 0. Hence there is a position with at least 6 matches. Call this the Yellow position, choose any 6 of the matched marked points and erase the other 9. Repeat the rotation process, but this time counting the number of matches of green points with the 6 marked points. The total number of matches is $6 \times 15 = 90$, so that the average number of matches per position is 2. As before, there is a position with at least 3 matches. Call this the Green position, choose any 3 of the matched marked points and erase the other 3. The 3 remaining marked points define three congruent triangles, a red one in the Red position, a yellow one in the Yellow position and a green one in the Green position.

Set S : Tournament Problems

Tournament problems are very popular in mathematics competitions because they are akin to each other. Arguably, the best mathematics competitions in the world is called the *International Mathematics Tournament of the Towns*, organized by a volunteer group of experts in Moscow under the leadership of the great **Nikolay Konstantinov**. We present three samples from this wonderful contest.

1. In a wrestling tournament, there are 100 participants, all of different strengths. Each wrestler participates in two matches. In each match, the stronger wrestler always wins. A wrestler who wins both matches is given an award. What is the least possible number of wrestlers who win awards?

2. In a tournament, each of 15 teams played each of the others exactly once. Prove that in at least one game, the sum of the number of games previously played by the two competing teams was odd.

3. In a tournament with no draws, each of eight teams plays every other team exactly once. Prove that at the conclusion of the tournament, there exist four teams A, B, C and D such that A has beaten B, C and D, B has beaten C and D, and C has beaten D.

Here are the solutions.

1. Let each wrestler be matched with the wrestler immediately below in strength, except for the weakest wrestler who is matched with the strongest one. Then each wrestler is in two matches. Everyone wins once and loses once except that the strongest one wins both matches while the weakest one loses both matches. Hence the smallest number of wrestlers who win awards is one, since the strongest wrestler always wins an award in any arrangement of matches.

2. In the 14 games played by each team, the 14 numbers of games played previously are 0, 1, 2, ..., 12, 13, respectively. Exactly 7 of these numbers are odd. Taking all the teams into consideration, exactly 105 of the 210 numbers of games played previously are odd. Hence it is impossible to form pairs with them, and at least one odd number will be matched with an even number, generating an odd sum for that game.

3. Among the eight teams, there must be one which has won no fewer games than any other team. Call this team A. A must have won at least four games. Consider the mini-tournament among the four teams beaten by A. There must also be one which has won no fewer games than the other three. Call this team B. B must have won at least two games in this mini-tournament. The two teams B has beaten play each other. Call the winner C and the loser D. Then we have four teams with the desired properties.

Set T : Packing and Covering Problems

Packing problems and their companions, covering problems, occur often in daily life. Our example to Problem 2 asks for the packing of the maximum number of S-tetrominoes into a 6 × 6 box. As we have seen, the answer is 8. What are the answers for the other tetrominoes?

For the O-tetromino, the answer is obviously 9. For each of the I-tetromino, the L-tetromino and the T-tetromino, the answer is 8. We first show, in Figure 18, that it is possible to pack 8 copies of each into a 6 × 6 box.

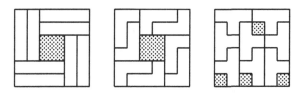

Figure 18

We now prove that these results are best possible.

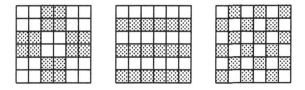

Figure 19

Suppose we pack copies of the I-tetromino into the box in Figure 19 on the left. Each copy must cover exactly 2 shaded squares. Since there are only 16 shaded squares, at most 8 copies can be packed.

Suppose we can pack 9 copies of the L-tetromino into the box in Figure 19 in the middle. Each copy either covers exactly 1 or 3 shaded squares. Hence 9 copies will cover an odd number of shaded squares. Since the number of shaded squares is 18, we have a contradiction.

Finally, suppose we can pack 9 copies of the T-tetromino into the box in Figure 19 on the right. Each copy either covers exactly 1 or 3 shaded squares. Hence 9 copies will cover an odd number of shaded squares. Since the number of shaded squares is 18, we again have a contradiction.

Interested readers may explore the packing into larger boxes, or the packing of larger polyominoes.

When the polyominoes are given uniform thickness, they become *poly-cubes*. For tetracubes, in addition to the five obtained from the tetrominoes, we have three other varieties, as shown in Figure 20. The two on the outside are left-handed and right-handed versions of the same shape.

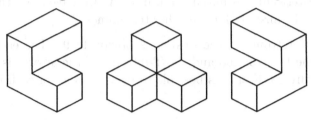

Figure 20

There are four other polycubes with up to four unit cubes that are not rectangular blocks. They are the V-tricube, the L-tetracube, the S-tetracube and the T-tetracube. Along with the three polycubes in Figure 20, their combined volume is 27. They can be packed into a $3 \times 3 \times 3$ box. This puzzle is the famous *Soma Cube*.

Set U : Dissection Problems

A key word in Problem 1 is *convex*. The general definition of a convex plane figure is such that whenever two points inside the figure are connected by a line segment, the entire line segment lies within the figure. In other words, every two people in a convex room can see each other.

For polygons, there is an alternative definition. A polygon is convex if and only if all of its interior angles are less than 180°. Thus every triangle is convex, but some quadrilaterals, such as the one shown in Figure 21 on the left, is not. Part of the segment joining two of its points lie outside the quadrilateral. One of its interior angles is greater than 180°.

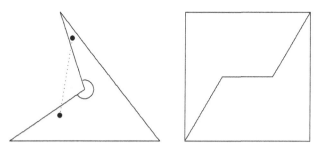

Figure 21

Problem 1 would have been trivial if we allow non-convex pentagons, as shown in Figure 21 on the right.

Problem 2 essentially asks for a dissection of a triangle into triangles. The triangulation process occurs very frequently in mathematics. An important case is when a convex polygon is dissected into triangles using only diagonals which do not cross inside the polygon. If the polygon is a quadrilateral, there are two ways as we can cut along either diagonal. If the polygon is a pentagon, there are five ways, as shown in Figure 22, even though they are all equivalent under rotation if the pentagon is regular.

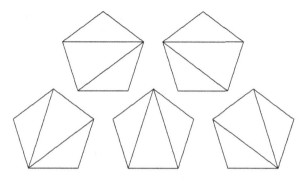

Figure 22

149

Take the middle triangulation in the bottom row of Figure 22. We label the edges of the pentagon other than the bottom edge w, x, y and z as shown in Figure 23 on the left. We have three triangles each with an unlabeled edge. Such an edge will be labeled by combining the labels of the other two sides in clockwise order. Figure 23 on the right shows the complete labeling.

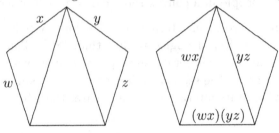

Figure 23

Just as there are five ways to triangulate a convex pentagon with non-intersecting diagonals, there are five ways to bracket the expression $wxyz$. Apart from $(wx)(yx)$, the other four ways are $w(x(yz))$, $((wx)y)z$, $w((xy)z)$ and $(w(xy))z$. They correspond in some order to the other four triangulations.

The phrase a "pretty little girls' school" has five different meanings according to how the words are bracketed.
(1) a very small school for girls;
(2) a beautiful school for small girls;
(3) a school for beautiful small girls;
(4) a beautiful small school for girls;
(5) a school for very small girls.

The number of ways of triangulating a convex polygon using only non-intersecting diagonals is given by a sequence called the *Catalan numbers*. The first few terms are 1, 1, 2, 5 and 14. The next term may be obtained by writing the sequence so far forward and backward, and adding up the pairwise products. The computation below shows that the next term is 42.

$$
\begin{array}{ccccccccc}
 & 1 & & 1 & & 2 & & 5 & & 14 \\
\times & 14 & & 5 & & 2 & & 1 & & 1 \\
\hline
 & 14 & + & 5 & + & 4 & + & 5 & + & 14 & = 42
\end{array}
$$

In Problem 3, all the pieces are identical. A covering of the plane by identical pieces is called a tessellation, and the pieces are called tiles. Every triangle can tile the plane because two copies of it can form a parallelogram, which tiles the plane. It turns out that every quadrilateral, whether convex or not, can also tile the plane. Such is not the case with convex pentagons. For instance, the regular pentagon clearly cannot tile the plane.

150

Five types of convex pentagons which can tile the plane were known as far back as 1918. Three more types were found in 1968, and it was believed that the list was complete. **Marjorie Rice**, an amateur mathematician, surprised the experts by finding a ninth type in 1977. Her work was made known to the professional mathematicians through the noted Escher-scholar **Doris Schattschneider**.

These nine types of pentagons abut edge to edge in their tessellations. In 1975, the first type of convex pentagons which tile the plane but not edge to edge was found. Marjorie Rice found three more of these types. The fifth type was found in 1985, and the sixth type was found in 2015. It is now believed that the list is complete with fifteen (nine plus six) types.

Unlike earlier types, the last type does not consist of a family of related pentagons but a single one. This is shown in Figure 24. The edge lengths are $AB = BC = DE = 1$, $CD = 2$ and EA just under 2. The angles are $\angle A = 105°$, $\angle B = 90°$, $\angle C = 150°$, $\angle D = 60°$ and $\angle E = 135°$.

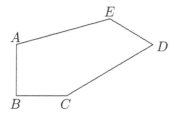

Figure 24

Apart from the square, the equilateral triangle, the regular hexagon is the only other regular polygon which tiles the plane. These three tessellations, shown in Figure 25, are called the Platonic tessellations. Each vertex is surrounded by the same sequence of identical polygons. They are (4,4,4,4), (3,3,3,3,3,3) and (6,6,6), respectively.

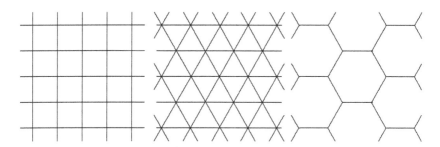

Figure 25

If we allow more than one kind of regular polygons but retain the requirement that all vertex sequences are the same, there are eight combinations called the Archimedean tessellations. They are shown in Figure 26. The vertex sequences are (3,3,3,3,6), (3,3,4,3,4), (3,3,3,4,4), (3,4,6,4), (3,6,3,6), (4,8,8), (4,6,12) and (3,12,12), respectively.

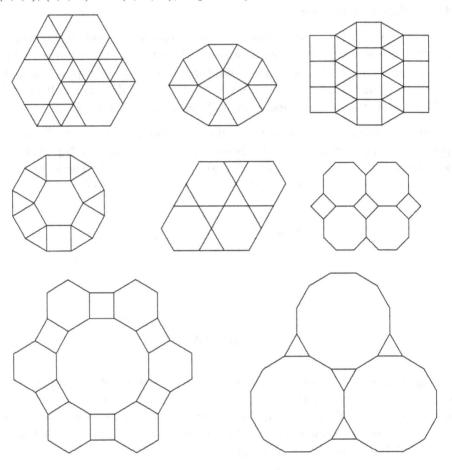

Figure 26

Set V : Graph Theory

A most useful concept in modern mathematics is that of a *graph*. It is simply a collection of points, called *vertices*, which are joined pairwise by lines, called *edges*. Sometimes, edges may have a specific direction.

This general definition gives graphs great versatility. Often, graphs are used as transportation models. Let the vertices represent cities and the edges represent roads connecting pairs of cities. If we travel from city to city by roads, we are moving along a *path* in the graph. It consists of a sequence of distinct edges such that two consecutive edges share a common vertex. If the path returns to its starting point, it is called a *cycle*

Let us reword the solution to Problem 1 in this new language. Once again, the cities are the vertices. From vertex A, draw a directed edge to vertex B if city B is the farthest from city A. Suppose we start from A, goes to B and comes back to A but not immediately. Then we have traveled along a directed cycle. However, in this cycle, each edge is longer than the preceding one, which is clearly impossible.

In Problem 2, a road connects more than two vertices. The underlying structure is no longer a graph but a *hypergraph*. The Fano plane we have encountered in Set D is such an example. It has seven vertices and seven hyper-edges, each connecting three vertices. It is redrawn as in Figure 27, where one of the hyper-edges is represented by a circle instead of a line segment.

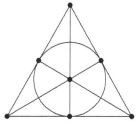

Figure 27

A special class of graphs consists of the complete graphs K_n, consisting of n vertices all pairs of which are joined by edges. Figure 28 shows K_4 and K_5.

Figure 28

Another special class of graphs are the complete bipartite graphs $K_{m,n}$, consisting of m vertices on one side and n vertices on the other. Every vertex on one side is joined to every vertex on the other, but no two vertices on the same side are joined. Figure 29 shows $K_{2,3}$ and $K_{3,3}$. Vertices on different sides are given different sizes.

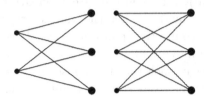

Figure 29

In all four graphs in Figures 28 and 29, edges intersect at points which are not vertices. If this can be avoided, the graph is said to be *planar*. Otherwise, it is said to be *non-planar*. It turns out that K_4 and $K_{2,3}$ are planar, while K_5 and $K_{3,3}$ are not. The minimum number of unavoidable crossing is called the *crossing number* of a graph. Thus a planar graph is one with crossing number 0, and Problem 3 shows the crossing number of K_6 is at most 3.

In Figure 30 on the left, we draw K_4 without crossing edges, dividing the plane into three triangles and an infinite region. To expand it to K_5, we must add a vertex to one of the four regions. It is easy to check that whichever our choice is, one of the other four vertices cannot be connected to the new vertex without crossing an edge.

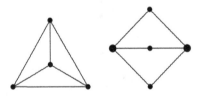

Figure 30

In Figure 30 on the right, we draw $K_{2,3}$ without crossing edges, dividing the plane into two quadrilaterals and an infinite region. To expand it to $K_{3,3}$, we must add a vertex marked with large circles to one of the three regions. It is easy to check that whichever our choice is, one of the three vertices marked with small circles cannot be connected to the new vertex without crossing an edge.

In puzzle books, $K_{3,3}$ is often called the utility graph. The three vertices on one side represent supplies of water, gas and electricity while the three other vertices represent feuding households who do not even want their supply lines to cross.

A theorem due to Kuratowski essentially states that a graph is planar if and only if it does not contain K_5 or $K_{3,3}$. To show that a graph is planar, it is often easier to redraw it without crossing edges. It is the converse of this result, incidentally the easier half, that is most useful. We give an example here.

A graph which appears very often is the Peterson graph, shown in Figure 31 on the left. It has ten vertices and fifteen edges. We now prove that it is not a planar graph. In Figure 31 on the right, we have deleted two edges and suppressed four vertices, leaving behind a copy of $K_{3,3}$. Again, vertices on different sides of the bipartite graph are given different sizes.

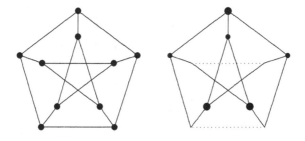

Figure 31

Set W : Numerical Solitaire Games

All four problems are solitaire games involving numbers. Here is an adventure in which the last two of the three parts are solitaire games involving numbers, though in a very different way.

Buck, Huck and Puck were magically transported into a dark room inside an old castle. They stumbled onto a robot and activated it. It said its name was Muck, and it had a lantern and a map of the castle. It consisted of 49 rooms in a 7 × 7 configuration, as shown in Figure 32. Muck informed them that their present location was in the room marked X. Most of the doors between adjacent rooms are one-way, indicated by arrows. The doors labeled with upper-case letters were locked. The keys with matching labels in lower-case letters were scattered among the rooms. With the correct key, a locked door could be opened from either side. How did they escape from the castle?

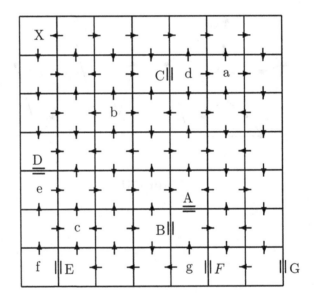

Figure 32

Once outside the castle, the party discovered that they were on top of a high cliff. The only means of descent was a pulley. Muck informed them that if the difference in weight between the two baskets was 5 kilograms or less, the heavier basket would descend at a rate which caused no harm to the occupants. If the difference was greater, the heavier basket would crash to the ground below. Now Buck weighed 65 kilograms, Huck 35 kilograms, Puck 30 kilograms while Muck weighed 25 kilograms. Moreover, Muck could withstand the crash. How did they come down from the top of the cliff?

Once they were on the beach, they discovered that they were on an island. The only means for getting back to the mainland was via a narrow bridge, just wide enough for any two of them to cross at the same time. By then, the sky was so pitch dark that Muck's lantern had to be carried while crossing the bridge. This meant that at least two people must cross to the mainland at the same time, so that one of them could bring the lantern back for anyone still on the island. They would cross at the slower speed of the two. Muck could cross the bridge in 1 minute, Buck in 2, Huck in 5 and Puck in 10. As soon as someone stepped on the bridge, a doomsday clock would be activated, and the bridge would self-destruct in 18 minutes. How did they escape back to the mainland?

The first two parts of this adventure are adapted from Boris Kordemsky's *Moscow Puzzles*, translated from Russian. It is a highly recommended mathematics book for children, now an inexpensive Dover paperback. The last part is from more recent mathematical folklore.

The maze with the one-way doors can be solved easily by picking up the keys in alphabetical order. To descend the cliff, Muck crashes down and is brought up by Puck, who is in turn brought up by Huck. Muck crashes down again, and is brought up along with Huck by Buck. The earlier moves are then repeated to get everybody down. To cross the bridge, Muck and Buck go first. Muck come back. Then Huck and Puck go, and Buck comes back. The last crossing is a repetition of the first one.

Set X : Geometric Solitaire Games

All four problems are solitaire games involving moving objects. Perhaps the most famous example of this type of puzzles is the Tower of Hanoi. There are three pegs in the playing board, and three disks of different sizes. The initial position is shown in Figure 33 on top, and the final position at the bottom. The rule is that we may only move a disk on top of a peg to the top of another peg, and a disk may not be placed on top of a smaller disk.

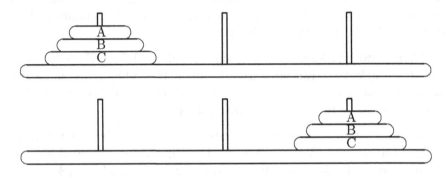

Figure 33

The task can be accomplished in seven moves. The readers are invited to discover how many moves are required if the number of disks is some number greater than three.

We now present here a delightful variation from Japan called the Arches of Hanoi. It consists of a base board with two pegs, and three arches, as shown in Figure 34.

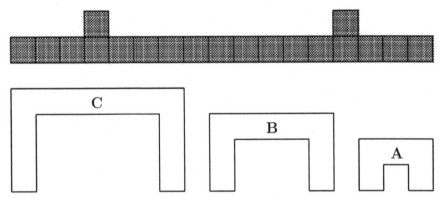

Figure 34

The initial position is shown in Figure 35 on top, and the final position at the bottom.

158

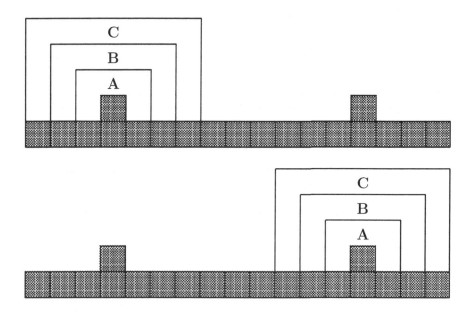

Figure 35

The arches may only be moved one at a time to anywhere on the base board. Except during movement, the bottom of both feet of each arch must lie flat directly on top of the base board. The task may be accomplished in 25 moves.

Set Y : Numerical Two-Player Games

All three problems are two-player games involving numbers. The most famous of such games is *Nim*. In the classic setting, we have three numbers 3, 5 and 7. Anna goes first, and turns alternate between her and Boris thereafter. In each turn, a player must reduce one of the non-zero numbers still remaining by an integral amount. The player who makes the last possible reduction is the winner.

We will analyze the game from Anna's point of view. The initial position is (3,5,7) and the final position is (0,0,0). If Anna can reach the final position, she wins. Thus we call (0,0,0) a safe position.

We will classify each position as either safe or unsafe. A position from which we can reach a safe position will be unsafe. All of the positions (0,0,1), (0,0,2), (0,0,3), (0,0,4), (0,0,5), (0,0,6) and (0,0,7) are unsafe, because if Anna leaves such a position to Boris, Boris will win by reducing the remaining number to 0.

A position from which every move leads to an unsafe position is safe. For instance, (0,1,1), (0,2,2), (0,3,3), (0,4,4) and (0,5,5) are all safe. From (0,1,1), Boris must go to the unsafe position (0,0,1). From (0,2,2), instead of going to the unsafe position (0,0,2), he can temporize by going to (0,1,2), but Anna can then leave him with the safe position (0,1,1). It is easy to see that (0,3,3), (0,4,4) and (0,5,5) are safe positions.

More generally, if only two equal numbers remain, the position is trivially safe. On the other hand, if two numbers are equal and the third number is non-zero, or if only two unequal numbers remain, the position is trivially unsafe. From the position (1,2,3), we can reach (0,1,2), (0,1,3), (0,2,3), (1,1,2), (1,1,3) or (1,2,2). All of these are trivially unsafe. Hence (1,2,3) is a safe position.

For the (3,5,7)-game, there are 88 different positions. However, there is no need to consider all of them. It turns out that Anna has three winning opening moves, reducing the position to one of (2,5,7), (3,4,7) or (3,5,6). The chart in Figure 36 shows her winning strategy if she leaves (3,5,6) for Boris. From each safe position, which is in boldface, we must consider all possible responses from Boris. From each unsafe position, we only need show a suitable response from Anna.

Interested readers may try to construct a similar chart if in her opening move, Anna leaves (2,5,7) or (3,4,7) for Boris.

Nim can be played with more than three positive integers of arbitrary size. Clearly, a case by case analysis as we have done so far would not work. A general method for the solution of *Nim* exists, making use of *binary numbers* or numbers in base 2.

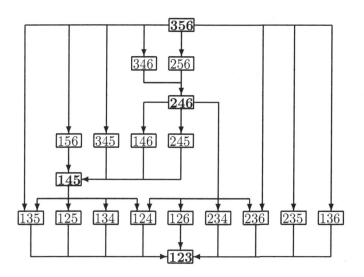

Figure 36

Set Z : Geometric Two-Player Games

The most popular two-player game with a geometric flavor is undoubtedly *Tic Tac Toe*. Anna and Boris take turns placing counters on a 3×3 board. Anna goes first and places red counters, while Boris places green counters. Whichever player first fills a row, a column or a diagonal with three counters of his or her color wins the game. Unlike *Nim* in the last section, this game can end in a draw. In fact, neither player has a winning strategy.

Our example to Problem 2 is a variant of this ancient game. Once again the goal is to achieve a certain configuration, but this time the objective is the S-tetromino. By using other polyominoes, we have a whole new class of games. We shall restrict our attention to square boards. Clearly, Anna has the advantage.

Consider the L-tetromino. On a 3×3 board, Anna cannot force a win since she cannot even fill a row or a column. On a 4×4 board, she has an easy win. In her first two moves, she can guarantee to get two adjacent squares among the four interior squares of the board. If Boris occupies a central square in his first move, Anna can get three adjacent red counters and win. If Boris plays on a border square in his first move, Anna will have an even easier time.

For the T-tetromino, Anna can force a win on a 5×5 board but not on a 4×4 board. For the I-tetromino, she can force a win on a 7×7 board but not on a 6×6 board. The readers can try to verify these results.

The most interesting case is the O-tetromino. It turns out that Boris can prevent Anna from winning, even on an infinite board! His simple strategy is to tile the plane with 1×2 rectangles, but in a brick-laying fashion as shown in Figure 37. Whenever Anna puts a red counter in a square, Boris puts a green counter in the other square of the brick. Since an O-tetromino must cover a complete brick, Anna cannot win!

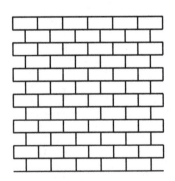

Figure 37

Short Answers

1979

    1. See Figure 1. **2.** 1957. **4(b).** 5. **5.** 35.

1980

    1. No. **2.** 3. **5.** See Figure 2.

1981

    2. No. **3.** See Figure 3. **4.** 72. **5.** Yes, 111111111.

1982

    1. 63. **2.** No. **5.** See Figure 4.

1983

    1. 23. **2.** See Figure 5. **3.** Liars: Benny and Kenny. **6.** 7, 4, 3 and 0.

1984

    2. See Figure 6.

1985

    4. 28, 38 and 998 copies of 1.

1986

    4. See Figure 7. **6(a).** 1369872. **6(b).** No.

1987

    2. No. **3.** See Figure 8. **4.** 1. **5.** 1001. **6.** 10.

1988

    1. No. **3.** 51, 1, 52, 2, 53, 3, ..., 99, 49, 100, 50. **4.** No. **5.** No. **6.** Yes.

1989

    1. 33. **5.** 166667 and 333334. **6.** Boris.

1990

    3. No. **4.** Anna.

1991

    2. No. **4.** 6. **5.** No. **6.** No.

1992

    2. 25.

© The Editor(s) (if applicable) and The Author(s), under exclusive license to Springer Nature Switzerland AG 2020
K. Garaschuk, A. Liu, *Grade Five Competition from the Leningrad Mathematical Olympiad*,
Problem Books in Mathematics, https://doi.org/10.1007/978-3-030-52946-8

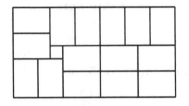

Figure 1

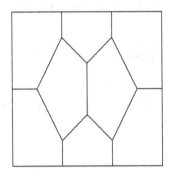

Figure 2

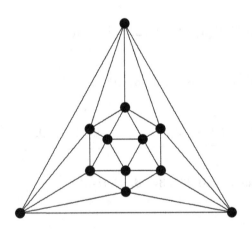

Figure 3

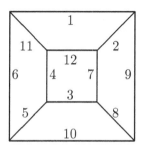

Figure 4

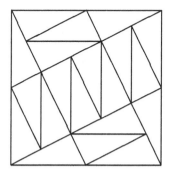

Figure 5

-1	2	-2	1
-2	1	-1	2
2	-1	1	-2
1	-2	2	-1

Figure 6

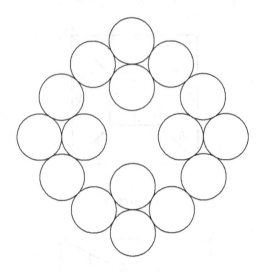

Figure 7

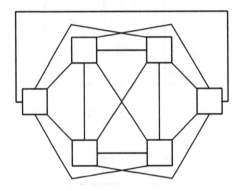

Figure 8

First Problem Index

Year	#1	#2	#3	#4	#5	#6
1979	T2	I3	C2	J2	F1	
1980	N1	F2	R1	P4	U1	V2
1981	G2	R2	U2	C3	K1	M2
1982	I1	B2	D2	H2	N3	Y1
1983	S1	U3	O1	W4	O3	J3
1984	I2	N2	B3	D3	W1	Y3
1985	P1	K2	V1	H3	C1	X3
1986	X1	Q2	H1	Q1	A2	J1
1987	W3	M1	V3	G1	L2	Z1
1988	W2	D1	A1	H4	X2	X4
1989	E1	L3	G3	P2	L1	Z2
1990	B1	P3	T3	Y2	A3	E3
1991	E2	F3	A4	O2	Z3	S3
1992	S2	Q3	K3	T1	R3	M3

Second Problem Index

Set	#1	#2	#3	#4
A	1988-3	1986-5	1990-5	1991-3
B	1990-1	1982-2	1984-3	
C	1985-5	1979-3	1981-4	
D	1988-2	1982-3	1984-4	
E	1989-1	1991-1	1990-6	
F	1979-5	1980-2	1991-2	
G	1987-4	1981-1	1989-3	
H	1986-3	1982-4	1985-4	1988-4
I	1982-1	1984-1	1979-2	
J	1986-6	1979-4	1983-6	
K	1981-5	1985-2	1992-3	
L	1989-5	1987-5	1989-2	
M	1987-2	1981-6	1992-6	
N	1980-1	1984-2	1982-5	
O	1983-3	1991-4	1983-5	
P	1985-1	1989-4	1990-2	1980-4
Q	1986-4	1986-2	1992-2	
R	1980-3	1981-2	1992-5	
S	1983-1	1992-1	1991-6	
T	1992-4	1979-1	1990-3	
U	1980-5	1981-3	1983-2	
V	1985-3	1980-6	1987-3	
W	1984-5	1988-1	1987-1	1983-4
X	1986-1	1988-5	1985-6	1988-6
Y	1982-6	1990-4	1984-6	
Z	1987-6	1989-6	1991-5	

© The Editor(s) (if applicable) and The Author(s), under exclusive license to Springer Nature Switzerland AG 2020
K. Garaschuk, A. Liu, *Grade Five Competition from the Leningrad Mathematical Olympiad*,
Problem Books in Mathematics, https://doi.org/10.1007/978-3-030-52946-8

Printed in the United States
by Baker & Taylor Publisher Services